Reliability of Nuclear Power Plants

Reliability of Multiphysical Systems Set

coordinated by
Abdelkhalak El Hami

Volume 14

Reliability of Nuclear Power Plants

Methods, Data and Applications

Edited by
André Lannoy

iSTE

WILEY

First published 2022 in Great Britain and the United States by ISTE Ltd and John Wiley & Sons, Inc.

ISTE Ltd
27-37 St George's Road
London SW19 4EU
UK

www.iste.co.uk

John Wiley & Sons, Inc.
111 River Street
Hoboken, NJ 07030
USA

www.wiley.com

Library of Congress Control Number: 2022938736

British Library Cataloguing-in-Publication Data
A CIP record for this book is available from the British Library
ISBN 978-1-78630-761-3

Contents

Foreword by Philippe Le Poac

The authors asked me to write this foreword in my capacity as President of the IMdR (*Institut pour la Maîtrise des Risques* [Institute for Risk Management]), but also perhaps in my capacity as someone who has spent 40 years of his professional life at CEA, *le Commissariat à l'Energie Atomique* (French Atomic Energy Commission), created on the order of General de Gaulle on the October 18, 1945, and which became the *Commissariat à l'Energie Atomique et aux Energies Alternatives* (Alternative Energies and Atomic Energy Commission) in 2010.

You must therefore forgive me if this foreword is strewn with personal recollections.

Operational safety and dependability are parts of the DNA of the IMdR. And operational safety and dependability include reliability, maintainability, availability and security.

The first chapter recalls the history of the notion of reliability and the word "reliability". There are domains where reliability is independent of security. We may even think of cases where systems had functionnally failed but were still safe as they had been safely switched off. One essential characteristic of the nuclear domain is that reliability is inseparable from security.

I am obliged to talk about vocabulary. For my entire professional life, like all nuclear professionals, I used these words, *sûreté nucléaire* an old translation of the English words, "nuclear safety".

Law 2006-686 of June 13, 2006 relating to Transparency and Security in the Nuclear field, called the TSN law, officially created the ASN (*Autorité de Sûreté*

Nucléaire [Nuclear Safety Authority]) and defines nuclear safety as the technical provisions and organizational measures relating to the design, construction, operation, halting and dismantling of static nuclear facilities, as well as to the transport of radioactive substances, taken with a view to preventing accidents or limiting their effects.

After I became President of the IMdR and was faced with other industrial terminologies, I noticed that the English word safety meant *sécurité*. Outside the nuclear domain, it is accepted that *sécurité* (safety) involves the protection of individuals, property and the environment from the risks that a system could create. And *sûreté* (security) involves the fight against any ill-intent that could threaten the system. The authors, aware of this linguistic problem, moreover, use both words security/safety in several places.

Chapter 3 presents the principles of calculating reliability in PSA, (Probabilistic Safety Assessment); at level 1, its main objective is to calculate the risk of meltdown. These PSAs rely on two types of model: fault trees and event trees.

The starting point for an event tree is what is called an initiating event. This may be a simple fault, such as a broken pipe, or the result of a fault tree. It is the same for events corresponding to the branching of event trees.

The representation in graph form of a fault tree and of event tree recall a "bowtie" currently used outside the field of nuclear science in studies on danger imposed by the European SEVESO directive for chemical, oil and gas facilities.

To construct event trees, the nuclear domain makes intense use of physical modeling and digital simulation, which has led to considerable progress because of the enormous increase in computer processing capacities.

Thus, the deterministic thermohydraulic code CATHARE, developed during a collaboration between CEA, EDF, AREVA and the IRSN, makes it possible, paired with a neutron code, to calculate precisely the evolution of temperature and pressure in the primary circuit in normal or accidental situations.

Reliability requires input data: feedback and expertise (Chapter 2). Operation feedback includes "event" and "material" feedback.

There are active components that are not reparable but which can be changed during corrective or preventive maintenance. There are also active components that are reparable. For these components, statistical data will make it possible to evaluate their reliability.

One particular case concerns active components on standby: they are not usually in use, but should be usable to ensure safety functions when needed. These components should be tested and failures during tests feed into the data needed to evaluate reliability.

Chapter 5 presents the main statistical and probabilistic modeling used for the reliability of industrial equipment, underlining the diversity of approaches due to their nature, the impact of maintenance, the quality of feedback and the complexity of the real world.

The effect of maintenance for reparable equipment is modeled, when the maintenance to be corrective, that is, carried out after a failure is detected and intended to return the reparable equipment to a state in which it can accomplish the required function, or when the maintenance to be preventive, that is, carried out at predetermined intervals or according to prescribed criteria and intended to reduce the likelihood of component failure.

It is noted that all the frequential approaches rely exclusively on the data available from the feedback: the quality of the results of the reliability study depends strongly on the quality of the data.

HUMS (Health and Usage Monitoring Systems) monitoring data are collected with the help of an increasing number of sensors with an increasingly tight time sampling. The data mass is becoming so large that the authors of this book speak of a transition from too little to too much and use one chapter (Chapter 7) to focus on methods developed to tackle the impact of big data on reliability. Many questions arise: it is a question of collecting relevant data and of being able to pre-treat and validate massive data. If machine learning techniques can help, there is no question of putting blind faith in black box models. Influencing factors, unsuspected at the outset, can be distinguished. But it will only be possible to have trust in big data algorithms if their results are explicable and explained. And the authors insist on the need for recourse to physical models and to analyses of organizational and human factors.

When the data are insufficient, Bayesian methods are needed, combined with expert opinion.

There are instances where feedback reveals no failures. This is the case when the equipment studied is very reliable or designed with a broad security margin.

It is also the case with a non-reparable passive component such as the reactor vessel.

Structural reliability is therefore necessary. This is the subject of Chapter 4.

Failure of a structure occurs when a stress written S exceeds a resistance written R. The stress-resistance method permits comparison with a numerical mechanical code for the structures. The stress and the resistance are expressed by variables that are considered random. The probability of failure is then determined by propagating the uncertainties in the physico-mathematical simulation code. The number of calculations using the Monte Carlo method becomes prohibitive for rare events and other methods are presented.

Mechanical tests make it possible to determine the material's resistance variables.

An important part of nuclear safety relies on the integrity of the vessel. The mechanical resistance of this thick metal structure has given rise to important research in fracture mechanics, the science of cracks propagation. The probabilistic approach to the risk of a sudden fracturing of the vessel consists of propagating uncertainties in the mechanical model of fracturing for various thermo-hydraulic transients. The toughness of the material, a measure of its resistance to crack propagation, is the first variable considered as random. Studies carried out by CEA, EDF and AREVA concluded that a probabilistic representation of resistance can be given by a three-parameter Weibull law.

I offer a personal aside. As a young engineer who had just left college, I discovered linear elastic fracture mechanics (LEFM) and the ASTM E399 standard (Standard Test Method for Linear-Elastic Plane-Strain Fracture Toughness of Metallic Materials) for determining the K_{IC}, the critical stress intensity factor, a measure of the material's resistance. I was one of the French researchers who developed Elasto Plastic Fracture Mechanics (EPFM) in France in the middle of the 1970s to extend Linear Elastic Fracture Mechanics and consider the plasticity of the steel from which the vessel is made. And I carried out, with others, the first research on the J-integral. Much later, I had the honor of directing the Department of Nuclear Materials at the Saclay CEA Center. Research on the J-integral resulted in the meantime in the ASTM E813 standard (test method for J_{IC}, a measure of fracture toughness).

Reliability security/safety is not only evaluated at the moment of design and implementation. They are evaluated periodically. Extending the lifetime of nuclear power plants necessitates studying the ageing of materials and more particularly the effects of radiation. Indeed, since extending lifetime amounts to reducing resistance or, in other words, increasing fragility, the evolution of this characteristic over time and as the cumulative effects of radiation have to be studied.

To do this, samples are irradiated in a test reactor or in a power generating reactor. It is then necessary to measure the properties of the irradiated samples in "hot labs", that is, in laboratories equipped with hot cells, able to receive highly radioactive materials in complete safety, and remote manipulators making it possible to proceed to experiments.

But the reliability and security/safety of complex sociotechnical systems is not only the concern of the engineering sciences. Human and social sciences also contribute. The accident at Three Mile Island in 1979, which occurred in the wake of inappropriate actions by the operators, has shown the importance of the human factor.

Chapter 6 addresses the human and organizational dimensions of reliability and nuclear safety.

The organizational dimension has been added because it is not only individual human error, which is always possible, that should be taken into account. The whole of the organism should be examined since reliability and security/safety are also dependent on management, who make decisions, arbitrate, allocate resources, manage skills organize feedback, and develop values and culture.

The TSN law cited above says that nuclear security includes nuclear safety, radioprotection and the prevention of and the fight against malicious activity, as well as civil security actions in case of accident.

Nuclear security therefore includes nuclear safety, but adds other considerations such as the fight against malicious activity.

A human, a friend of the system, can commit an involuntary error but there may also be other humans, enemies of the system, who may seek to harm the system voluntarily. Reliability and security/safety also depend on the lines of defense, both technical and organizational, that will have been planned against malicious attacks, whether physical or based on information technology.

The present work considers reliability studies carried out in the domain of nuclear electricity production but some approaches and methods may be applicable to other industrial sectors. Society, which has always been demanding over nuclear power considering the risks it runs, is becoming more so with other economic activities. Experience acquired in the nuclear domain could usefully inspire those in charge of reliability, operational safety and complex sociotechnical security systems.

Philippe LE POAC
President of the IMdR

Foreword by Antoine Grall

Society is attentive to risks for individuals and the environment, and faced with ever more sophisticated production systems and subject to the goals of optimization and cost management, operational safety is taking an ever greater place. Many unwanted events can, when they are considered individually, be evaluated as very unlikely and inconsequential. Nevertheless, within complex systems, cumulative effects and the effects of interactions can be very significant. An operational safety approach does not aim to eliminate all risk, but rather to keep it under control. This means determining and characterizing a design and an operational mode allowing a good balance between different criteria linked to operation of the system and to its interaction with the environment. The exact characterization of the "right" balance can be highly variable from one application domain to the other depending on the importance given to economic, reliability and environmental aspects, etc.

This book focuses on electricity production systems and on the nuclear domain. "Reliability" and "nuclear energy" are terms that naturally go together. No one can deny the important role of the progress undergone in methods and studies on operational safety in the context of work carried out in the sector of nuclear power production. This book, coordinated by André Lannoy, exclusively brings together contributions from authors with recognized competencies in reliability and significant experience in the nuclear field. It demonstrates the broad spectrum and complexity of the questions tackled.

Should we believe that there is a reliability specific to the nuclear domain? In other words, is it relevant for the reader to immerse themselves in the different chapters if they are interested in reliability in association with other application domains? The answer, in my opinion, is definitely affirmative. Independently of the subject studied, any operational safety engineer has the same set of techniques and methods on which they can call. Their work involves using these as well as possibly starting from a refined understanding of the dysfunctional mechanisms at play. A nuclear energy production plant represents a complex system, within which various issues co-exist, transverse to many domains, linked for example to the analysis of passive components (structures), dynamic systems (active components and control-command), human performance, etc. In the field of nuclear safety, there is a continuous demand for improvement that leads to periodic re-examinations and to consideration of the evolution of knowledge. Points to emphasize, procedures and the regulatory framework may be specific but junior and confirmed senior specialists will be able to draw significant benefit from the shared expertise offered by the authors of the different chapters.

Many aspects are addressed throughout these pages and the frameworks for reading may be diverse. The data are present at every stage of the reliability studies. Whether it comes from accelerated testing, from expertise, from operational experience feedback or from sensors in real time, it is vital to use it. Three chapters are particularly associated with data. Chapter 2 provides a general perspective on feedback and the associated concepts, highlighting contributions from events and from materials. Chapter 5 gives the reader an overall vision of the main stochastic modeling frameworks based on the data and used in reliability. They have the aim of describing the behavior of specific sub-systems or components and in practice are chosen, adjusted and validated through confrontation with the data observed. Finally, Chapter 7 is directed toward a current issue, of collecting and processing the massive surveillance and monitoring data. This new challenge has emerged in particular because of modern monitoring devices. It is a particular challenge for the field of reliability, which has historically had to deal with little data.

Monitoring and feedback data are clearly not the only source of information available on the state and behavior of a system or a component. As for passive components in particular, physical or mechanical behavioral models are available. Analyses of the reliability of structures represent one particularly interesting example of association between these deterministic behavioral models and uncertainties on the variables. The methodological framework is presented in Chapter 4 and applied to examples.

An electricity production system based on nuclear energy is a complex system and two chapters focus on two particular aspects of this complexity. The first aspect tackles probabilistic studies in safety. These studies encompass very large models linking fault trees and event trees. Given the large number of components interacting, combinational problems are often present. In this context, it is particularly interesting to understand practical application. More recent approaches are also presented. The second aspect of complexity evoked hinges on the human and organizational dimensions that require a global and systemic vision. In particular, they mobilize competencies from the field of human and social sciences and demonstrate how much reliability is destined to be an interdisciplinary field.

The question of making human performance reliable is discussed with its different aspects.

To conclude, I would like to emphasize the benefits of this book for anyone interested in the field of reliability. As a teacher to future researchers, I can only recommend that students read it. Reading it may also make it possible to acquire a broad view of the many aspects of reliability as well as to complete, very favorably, an essential theoretical initiation that can be followed in an academic framework. Through clear and precise presentations, the book carries the dimension of applicability on real systems of methods that are often presented with simple illustrative examples.

As a researcher who cares about the applicability of his work, it seems helpful to me to be attentive to the perspectives developed and to the avenues for R&D mentioned by the various authors. They are linked to methodological obstacles that we must succeed in lifting and they give precise indications as to the hypotheses present in the existing theoretical frameworks that are the most restrictive.

Professor Antoine GRALL
Technical University of Troyes
Deputy director of the doctoral school
"*Science for the Engineer*"
May 2022

Preface

The authors of this edited collection wish to honor Henri Procaccia, a former expert research engineer, former group leader and former deputy head of department at EDF R&D, cofounder, honorary member and member of the ESReDA (European Safety, Reliability & Data Association) governing committee, member of the IMdR (*Institut pour la Maîtrise des Risques* [Institute for Risk Management]), for his pioneering role in the field of reliability in nuclear power.

Henri Procaccia was one of the active players in reliability and we are indebted to him for a great deal of research addressing this subject in depth. From the start of the 1970s, well before the coupling of the first pressurized water reactor (using US technology) in France at Fessenheim, Henri Procaccia worked on system reliability and on the structuring of operation feedback for pressurized water reactors and fast neutron reactors.

In the 1980s, the low quantity of feedback data issuing from power stations that were still at their initial stage led him to adhere to the Bayesian approach, which he promoted both in his teaching and practice. He was responsible for the first data handbook on reliability, the SRDF (Collection and Analysis System for Reliability Data). At the end of the 1980s, on the basis of his skill in physical testing, he convinced the EDF to engage in analyzing structural reliability and in fatigue monitoring.

In the 1990s, his field of investigation primarily focused on optimizing maintenance and on decision support. In all these topics, reliability emerged as an essential parameter. The authors are grateful to Henri Procaccia for having thus aided the development of reliability methods, their application and their use. He was also motivated by a profound sense of humanity.

May 2022

Acknowledgments

The authors of this book would warmly like to thank Monsieur Abdelkhalak El Hami, Professor at INSA, Rouen and Director of the Department of Mechanics and "Industrial performance and innovation", for suggesting that we write a state of the art of knowledge on the theme of reliability in nuclear power plants. We are indebted to him and his services for technical support in creating this book.

First of all we thank Mohamed Eid, former engineer-researcher at CEA Saclay, teacher and now consultant and President of the ESReDA, for his encouragement, advice, especially on methods, support with which he has been so generous, precise and perspicacious reviewing of this book, and so for the essential contribution due to which we have been able to complete this book.

We must also thank:

– Hervé Boll, Deputy Director of Programmes at EDF R&D, for reviewing Chapters 3 to 5 on systems, structures and components.

– Vincent Sorel, Project Manager, *Sûreté Nouveau Nucléaire* (New Nuclear Safety) and PSA expert, Cécile Luzoir, director of *EPS Agressions* (PSA aggressions), both at the EDF Technical Department, and Julien Tissot, head of the Group "EPS et disponibilité des systèmes" (PSA and system availability) in the "PERICLES" department at EDF R&D, for reviewing Chapter 3 of this book.

– Philippe Bryla, Mechanics and Materials expert at the EDF Hydro General Technical Department, Group C2M, for reviewing part of Chapter 4.

– Vincent Chabridon and Roman Sueur, research engineers at the PRISME (Performance, Industriel Risk and Monitoring for Maintenance and Operation) department at EDF R&D, for their relevant and constructive comments on Chapter 5.

– Franck Anner, at IRSN, head of the human and organizational expertise office, Sarah Fourgeaud, head of the Homme, Organisation, Technologie (Man, Organization, Technology) Service, Olivier Dubois, Deputy-Director of safety expertise at REP, and at CEA, Didier Balestrieri, Head of the *Service de Soutien et de Gestion de Crise* (SSGC) (Crisis Support and Management Service) at the *Direction de la Sécurité et de la Sûreté Nucléaire* (DSSN) (Department of Security and Nuclear Safety) at CEA, for their reviews on Chapter 6.

These comments and conversations and these varied experiences have enriched our thinking, and have enabled us to complete our approach to reliability and nuclear safety.

Thanks to Alexandra Jordan who carried out the work of translating this book into English, while providing us with her insightful and precise insights and suggestions.

Thank you again.

May 2022

Author Biographies

Emmanuel Ardillon, graduated from the École Nationale des Ponts et Chaussées (1990), holder of an M.Phil in "Probabilities and Applications" from the University of Paris VI, is a research engineer at EDF R&D (PRISME Department) in the field of reliability of structures, probabilistic approaches to physical phenomena and uncertainty processing. He is the author of structural reliability analyses (methodology, applications) in the nuclear and hydraulic field, research project manager, editor of the collective work *Structural Reliability Analyses into System Risk Assessment* (ESReDA, 2010), contributor to the collective work *La fiabilité en mécanique – Des méthodes aux applications* (Presses des Mines, 2018) and leader of the working group "structural reliability and safety" of the Institut pour la Maîtrise des Risques (IMdR).

Marc Bouissou, graduate of the École Nationale Supérieure des Mines de Paris (1980), holder of an ability to conduct research from the University of Science and Technology of Lille and qualified for the post of University Professor in 2008, is a senior research engineer at EDF R&D (PERICLES Department). He has devoted most of his career to the development of innovative methods and tools for the automation of system dependability studies. He has more than 100 publications to his credit in this field, including several book chapters, and he has written the book "*Gestion de la complexité dans les études quantitatives de sûreté de fonctionnement de systèmes*" (Lavoisier, 2008). He has given numerous courses to master's students on systems reliability, in particular when he was a part-time professor at the École Centrale Paris from 2009 to 2015. Since its creation, he has been leading the working IMdR's group "methodological research".

Nicolas Dechy, graduate of the École des Mines de Douai (1999), is a research engineer at the Institut de Radioprotection et de Sûreté Nucléaire (IRSN) within the Human Organization and Technology Service of the Nuclear Safety Pole, where he has been carrying out expertise since 2010 as a specialist in organizational and human factors, and where he assesses risk management by operators. Since 1999, in a design office (AINF), then at INERIS, he has carried out studies, expertise and research on risk management, in particular on experience feedback, accident analysis, crisis, management maintenance and subcontracting. He contributes to the reflections, exchanges and activities of the associations IMdR, ESReDA, Institut pour une Culture de Sécurité Industrielle-Fondation pour une Culture de Sécurité Industrielle (ICSI-FonCSI) and Collectif Heuristique pour l'Analyse Organisationnelle de Sécurité (CHAOS).

Yves Dien began his career at EDF R&D in the early 1980s, notably through his participation in a design and ergonomic assessment project for a fully computerized control room for nuclear power plants. He was also involved in the design and evaluation of assistance to nuclear power plant operators (incident and accident control procedures, computerized tools in conventional control rooms, etc.). Then, for 10 years, he was responsible for nuclear affairs for the countries of Central and Eastern Europe (European TACIS and PHARE funds, bilateral conventions). Finally, back at EDF R&D, he oversaw a project on the organizational factors of incidents, accidents and crises. He contributed to the discussions and reflections of the IMdR and ESReDA associations, as well as the ICSI-FonCSI. Currently retired, he is a member of the CHAOS association which conducts research on the organizational factors of industrial security.

Mohamed Eid graduated in Nuclear Energy Engineering from the Department of Nuclear Engineering of the University of Alexandria (Egypt) in 1977, then Doctor of Engineering in Nuclear Reactor Physics from the University of Paris-XI in 1985; he joined the CEA as a research engineer where he developed his multidisciplinary expertise in neutronics, thermo-hydraulics, radiation protection and probabilistic risk analysis. He was in charge of continuing education courses in simulation by the Monte Carlo method and its applications in dependability since 1987 in several engineering schools: École Centrale Paris, École Nationale Supérieure de Techniques Avancées and Supaero. He also worked as an Associate Professor at the Institut National des Sciences Appliquées in Rouen between 2005 and 2020. He has published around 50 scientific articles and co-authored several books in English. He is a member of several editorial boards of scientific journals. He is an active member of European Safety and Reliability Association (ESRA) and IMdR and President of ESReDA. After his retirement from CEA in 2020, he founded his consulting and studies firm in Risk Modeling & Analysis "RiskLyse". Within the frame of this book, he advised all the authors and re-read several chapters, in particular Chapters 1, 3, 5.

André Lannoy is a research engineer, doctor of detonics, previously scientific advisor to EDF R&D, author or co-author of numerous books and communications, moderator of the IMdR product commission (projects, scientific days, training), member of the working groups “feedback from technical experience” and “structural reliability and safety” from IMdR and honorary member of ESReDA”.

Emmanuel Remy, graduate of the Institute of Statistics of the University of Paris (2001), is an expert research engineer at EDF R&D (PRISME, Performance, Industrial Risk and Monitoring for Maintenance and Operation) in the field of probabilities and statistical methods for equipment of nuclear, hydraulic and wind power generation. In addition to his applied research (in close collaboration with the academic world), and teaching activities, he has contributed to the group “reliability and uncertainties” of the French Statistical Society since its creation in 2009 and, since 2014, he has co-coordinated the working group “feedback from technical experience” from the IMdR”.

Jean-François Vautier is a specialist in Organizational and Human Factors at the CEA. For 20 years, he has been running the skills center and the network there. Since 2013, he has also been one of the facilitators of the working group “organization and risk management”. For the Technical Engineering Editions, he directs a collection of collective works that follow on from the working intergroup organized by the IMdR, as well as a collection on complex systems. He has published articles on Systemics in addition to others on organizational and human factors.

1

Aims and Introduction

This introductory chapter discusses the aim of this book, which is to explain the methods and reliability studies employed in the nuclear power production sector. It is not a book on advanced reliability but rather a guide to good practice in current use, in which it is permitted to propose R&D actions to make further progress. One section retraces very briefly the evolution of ideas and methods in the field of reliability since earlier times. With the design and implementation of nuclear power plants in the 1950s, then their operation and maintenance, concern about reliability have become very significant for the challenges of safety and performance. Methods, tools and knowledge have moreover greatly evolved since the 1970s. The content of this book is presented at the end of this chapter.

1.1. The aims of this work

This edited collection follows a suggestion by the editor in chief of a collected series of books on "applied reliability" on behalf of iSTE and Wiley & Sons editions, Professor Abdelkhalac El Hami, Director of the Mechanics and Performance Department at INSA, Rouen, Normandy (*Institut National des Sciences Appliquées;* National Institute of Applied Sciences). The collected series of books concerns reliability as it is practiced, estimated, approached, interpreted and integrated into the various industrial sectors (aero-space, car-making, nuclear, oil and gas, etc.).

The present work emphasizes *reliability studies* carried out *in the nuclear energy production sector*, given that the approaches, methods and results described in it can be applied to other domains such as hydraulic production, wind power, etc., but also

---

Chapter written by André LANNOY.

to other sectors of processing industries or transport, such as offshore platforms or rail transport, and so on.

Even if these approaches and methods are clearly generic, the reader who uses them must nevertheless check that the context is appropriate and the operational, maintenance and environmental conditions are substantially similar. This potential use must also conform to regulatory context of the industrial sector.

This book is not a piece of academic or advanced research on reliability. It sketches the state of the art on industrial practice on reliability to date. It presents the main approaches and methods used at the present moment in the sector of nuclear energy production. In this sense, it can be considered as a guide to good practice or rather as a vade mecum that can easily be consulted. The authors however do not refrain from listing views of progress and R&D, or indeed from proposing new developments.

1.2. Reliability, an application of probability theory

1.2.1. *What is reliability?*

The definition given by standard EN 13306 (edition 2, 2010) is as follows: *aptitude d'un bien à accomplir une fonction requise, dans des conditions données, durant un intervalle de temps donné (the ability of an item to perform a required function under given conditions for a given time interval).*

The first remark is that the standard refers to an "item". This term may not be unanimously accepted in the reliability community. Although it satisfies working reliability specialists, design reliability specialists or modelers might have preferred the term "functional entity".

The second remark to be made is that reliability is first of all a qualitative notion. It must be verified that the item concerned really guarantees all its functions considered necessary to provide a given service.

The third remark is that reliability lends itself well to the use of a probabilistic metric; it designates the value of the probability of something being in good working order during a well-defined interval of time. Reliability can therefore be quantified in the form of a probability or a probabilistic performance indicator. From 1969, Kaufmann presented it as a confidence indicator, which he called the technical confidence [KAU 69]. Lemaire [LEM 05] refers to the definition taken from standard AFNOR NF X50-120 from 1988: "the ability of an item to accomplish a required function in given conditions, over a given duration… the term is also used as a characteristic designating a probability of success or a percentage of success".

Another quantitative characteristic of reliability is the instantaneous failure rate λ (t), which designates the proportion, per unit of time, of devices that, once they have survived to an arbitrary instant, t, are no longer "living" at instant $t + dt$. $\lambda(t)$ is a reliability indicator indicating imminent failure at instant t.

We can distinguish several reliabilities depending on the phases of the lifecycle of an item [LAN 06]:

– comparative reliability, calculated from past feedback data, used in preliminary studies in the pre-project phase;

– allocated reliability, requirement, reference threshold value, defined in the specification phase;

– planned or theoretical reliability, calculated during design, and compared to the allocated reliability;

– practical reliability or operational reliability or measured reliability, estimated from operational feedback, when the equipment is in use;

– predictive reliability, anticipated reliability or extrapolated reliability or reliability prognosis, estimated from an operational feedback analysis at a given instant and considering contexts and future conditions of use.

1.2.2. *The early days of reliability*

Security and safety have always been pressing needs for societies [LAN 18]. Since the 3rd millennium BC, the resistance of houses to the threat of collapse was a great worry for Cycladic civilizations. Around 1730 BC, the sixth king of Babylon, Hammurabi, developed one of the first regulations intended to guarantee a level of security in homes (articles 229 and 230 of the code of Hammurabi, the Louvre). This is the first legal code. It remained in force for a millennium in Mesopotamia [LEM 05].

Vitruvius (1st-century BC) appears to be the *first proponent of reliability* in history (*De architectura*). He seems to have been the first to be concerned with dependability:

– needs and utility (*utilitas*) (he was the first to carry out a functional analysis, and we recall that any study of reliability, whatever it is, begins with a functional analysis);

– reliability, durability and the robustness of the item (*firmitas*);

– capitalization of existing knowledge, the codification of constructive provisions, resistance to natural harms;

– esthetics (*venustas*).

The word *fiabilité* (*reliability*) appeared in the course of the second half of the 13th century. In the 20th century, it became a very commonly use word for mastering technological risk. It comes from the Latin *fides*, faith, confidence, which produces confidence, good faith, and from the Latin *fido*, to trust, to have confidence in, to count on.

It was only in the wake of "the geometry of chance", begun in 1654 by Blaise Pascal, that the first stage of *probabilistic studies* focused on the risks caused by gambling and led to the establishment of mortality statistics. At the start of the 18th century, Daniel Bernoulli [BER 38] defined risk and considered it as a two-dimensional measure: "the probability or likelihood of an event" and the gravity of its consequences.

The posthumous work by Thomas Bayes [BAY 62], *An essay toward solving a problem in the doctrine of chance*, was published in 1763. This work would revolutionize reliability studies in the 20th century. The Bayesian method has now become unavoidable.

In 1776, Buffon [BUF 76] observed that a guild, the carpenters had a very imperfect knowledge of force and resistance of wood, which depended on the organization of the wood and on particular circumstances. Over several years, he carried out tests on wooden beams (he recorded: length, breadth, mass, load, breaking time, strain). These were, to our knowledge, the first mechanical reliability tests. At this time, regression methods did not exist, so it was not possible to develop a law governing lifetime. Reliability can therefore also be obtained experimentally.

And it was only after the 19th-century railways and then from the 1930s that reliability studies developed from needs linked to the aeronautics and electrical industries to civil engineering structures. We note Max Mayer's [MAY 26] proposition to use average values and parameter variations in design.

Between 1930 and 1940, Sir Alfred Grenville Pugsley expressed the first quantified risk objectives. He demanded that an airplane's accident rate "considering all the causes of breakdown likely to cause to an accident" should not exceed 10^{-5} per hour, including 10^{-7} per hour for causes linked to the airplane's structure [VIL 88]. Pugsley also applied probabilistic objectives to shape civil engineering at the end of the 1940s.

1.2.3. *The birth of modern reliability*

This period extended from 1937 to 1948 and the war years would create a considerable boom in reliability research.

In 1937, Bruno De Finetti [DEF 37] became pre-eminent in "subjective operational" probability design.

In 1939, the Swedish engineer and mathematician Weibull published his work on the distribution known as Weibull distribution, used for mechanical tests, especially fatigue tests [WEI 39]. Now, it is also much used in probability. It can be noted that Weibull thought that his law could not be used for reliability. It was used for this purpose, however, after the War. Other distributions would then be used in the field of reliability.

The Second World War then began. Industry had to deal with intense production for military ends. And the idea that production chains are as reliable as their weakest link came into question. Collaborating with Wernher Von Braun on V1 manufacturing chains, Eric Pieruschka, in collaboration with Robert Lusser, perfected the calculation of a chain's reliability as being the product of the probabilities of survival of each of the chain's components. Operational research appeared in 1940 in England, and then in the United States for military ends: for the United Kingdom, it was a matter of best using its military means, which were insufficient at the time (airplanes, anti-aircraft forces, naval methods). The fundamental idea was to take as much care as possible in using the means available for design and construction.

1.2.4. *The development of modern reliability 1948–1960*

In 1949 Murphy's law was announced, perhaps better known by the name of the "maximum poisoning law" or "buttered toast law": as soon as it is possible for anything to go wrong, it will! A law never demonstrated but always experimentally verified.

At the start of the 1950s, Epstein and Sobel [EPS 53] would present their work on the different distribution forms of a material's lifespan. The exponential law of lifetimes corresponds to the particular instance of the hypothesis of a constant instantaneous failure rate, particularly well adapted to electronic equipment, and very convenient for modeling. The use of Weibull's law would come to the foreground in the 1990s, with studies on ageing.

Moreover, the threats of the Cold War would, on the one hand, increase the development of effective arms systems for dissuasion and, on the other hand, create a lively emulation between the United States and the USSR over questions of national prestige. The fallout from military and space technology for telecommunications was plentiful (intelligence satellites for transmitting images, geolocation, mobile data, mobile telephony, etc.).

The US army developed FMEA, then FMECA. Military reference Mil-P-1629, entitled "Procedures for Performing a Failure Mode, Effects and Criticality Analysis", is dated November 9, 1949. This method was employed as a technique for evaluating failures to determine the reliability of a piece of equipment and a system, especially electronic systems. This method is now universally used in all sectors. It was developed by the aviation industry and the US army, with the aim of making reliable and securing aerospace flying equipment. Its use was then extended to industry to optimize the operational safety of components and systems representing a significant risk for individuals and the environment. Three concepts would come to complete that of reliability: availability, maintainability and support logistics, thus expressing the dependability.

The first studies on the human factor were launched in the United States.

The AIEE (American Institute of Electrical Engineers), became the IEEE (Institute of Electrical and Electronics Engineers) in 1963, and in 1954 held its first conference on reliability. It also saw the first use of Markov processes in industry.

In 1955, the first reliability study was published in France by CNET (National Center for Telecommunications Studies), the pioneer of reliability studies. It was then that the French word "*fiabilité*" (reliability) appeared, under the leadership of Paul Blanquart and Guy Peyrache [LAN 08]. The Académie Française went on to accept the term "*fiabilité*" in October 1965.

1.2.5. *The advent of reliability specialists 1960–1974*

With the design and implementation of nuclear power plants, security/safety concerns became significant. Potential accidents are classified, placed in a hierarchy and analyzed in detail, as much for their seriousness as for their frequency. So was born the defense in depth principle in the nuclear industry. The development of other types of industries (car-making, chemical, oil and gas, etc.) will take advantage of this impulse.

These advances therefore concern methodological progress: knowledge, approaches, methods, etc., but also very many advances for the whole of society (and considerable for industrial societies): the increase in lifetime, industries and safer transport, scientific progress, technical progress, greater security/safety linked to better prevention and protection policies, the approval needed for industries involving hazardous processes (the aim being to limit the potential harmful impacts from the design stage), quality approaches, actions and demonstrations by consumers that became widespread in the period between 1970 and 1976 for the defense of consumers rights and a better quality of products and food (Ralph Nader), etc.

Barlow and Proshan [BAR 65] show the impact of maintenance on reliability. It is only possible to calculate the reliability of an item if we estimate the effects of maintenance.

The very well-known Mil-Hdbk-217 (edition B) on electronic reliability was published in 1974.

Franck Reginald Farmer [FAR 67] was interested in the acceptability of risks and tackled the question of the maximal credible accident. From his point of view, all potentially dangerous events should be considered. The risk is represented by a curve called fC, frequency–consequences, showing at the same time events with a low probability and severe consequences and frequent events with low-level consequences. This notion was introduced in 1967 for iodine 131 fallout from nuclear reactors. We therefore find fC curves and FN (frequency–number of fatalities) curves traced in a log–log plot. This plot is traced during most risk analyses because of its simple and informative character. We therefore speak of the Farmer diagram curve or plot that considers three domains: (i) events with high frequency and high severity, (ii) events with average or median frequency, but substantial severity, and (iii) events with low frequency and extreme seriousness (the case of a major accident). These notions were reprised more recently, with [TAB 10, LEM 14] distinguishing the medianistan (the first two domains, where the probability can be estimated by frequential or Bayesian methods), the third domain, called extremistan where the decision will be made with regard to the severity of the major accident. These are the first steps in the probabilistic approach.

We also cite the first treaties on reliability in French: the translation into French of Bazovsky [BAZ 66], Chapouille and De Pazzis [CHA 68], Schwob-Peyrache [SCH 69], and Ligeron-Marcovici [LIG 74].

1.2.6. *The "safety culture decade" 1975–1990*

Studies on nuclear safety are intensifying because of the report by Rasmussen [USN 75] where the concept of a fault tree (for causes) and the event tree (for the consequences) appeared. These methods would also be used to study the Canvey Island oil complex (1978), which also saw the development of the use of physical models. These were the first safety evaluations published, deterministic and probabilistic. These analyses show the strategic benefit of operational feedback that was beginning to develop in most industries. Operation feedback is in fact the base material, the input data for all risk and dependability studies and for controlling risks.

Human factors, the importance of which was signaled in Wash 1400, have emerged in French nuclear safety institutions shortly after the accident at the nuclear power plant at Three Mile Island. Several analyses of this accident highlighted the importance of the role of humans as one of the essential links in safety. There then appeared ergonomists and specialists in human sciences, nuclear safety and human factor. The best-known work [SWA 83] is a fundamental reference for all later works. The methodology called THERP (Technology for Human Error Rate Prediction) estimates a probability of human error (which can be defined by a human action that has the potential to degrade a system or overwhelm a situation). Human is considered as one of the system components. These data are used universally in probabilistic safety evaluations.

We can point out the publication of the book by Michel Gondran and Alain Pagès on the reliability of systems in the 1980 [GON 80] and the publication of the synthesis work by Alain Villemeur in 1988 [VIL 88] on dependability methods in industrial systems.

1.2.7. *Maximizing efficiency, performances and profits 1990–2007*

The leitmotiv to which company managers referred at this time was clearly "still as good or better and still cheaper". This was about meeting the expectations of the triumphant market economy, always growing, and the quality of services and products. It was a constant quest for better efficiency, for better performances, for greater profitability. The "bare necessity" sought in design was then transposed, especially in maintenance. The RCM (reliability centered maintenance) method is developed and applied, on the one hand, to lower the costs of preventive maintenance and increasing availability and profitability, but also to better ensure the safety of industrial systems. At the end of the 1990s, many methods were devoted to the study of ageing in systems, structures and components (SSC) taking account of the amount of capital invested in nuclear power, refineries or in air, rail and maritime transport. New concepts appeared with lifecycle management (LCM). Re-design, replacements, refurbishment, maintenance programs and investments were optimized by asset management (AM). These studies were the first acts of predictive maintenance.

Moreover, since existing methods making it possible to make provisional reliability calculations for electronic components had not been updated for several years, various manufacturers desired a new methodology that would be both precise and realistic. It was in 1999 that a study named FIDES was specified, which would be developed over 3 years to become a reference work from 2005.

This period was also the period of the Gulf War in Kuwait. The world noted the logistical capacities of the US army. Logistical support suddenly became a considerable industrial challenge and design processes integrated logistical support, such as maintenance, which was optimized from the beginning.

The works of Reason [REA 97] establish a classification of human errors by distinguishing errors called active errors made by the operator from latent errors linked to the organization. These works would highlight the importance of organizational factors, which came to complete human factors (Organizational and Human Factors).

This period was marked by the success of entry into the year 2000, showing the usefulness and effectiveness of dependability methods.

1.2.8. *The return to safety, risk aversion 2007–2020*

This period was marked by an ever greater aversion to risk. Already at the start of the 2000s, events had occurred that were out of the ordinary: the attacks of September 11, 2001, the tsunami of 2004 in south-west Asia, etc. The general public lost confidence in the competence, capacities and seriousness of scientists, experts, politicians, economists, etc., to overcome unthinkable events and made them all responsible for these catastrophes. Distrust could be seen between the general public and this intellectual elite that made decisions on major projects. The aversion increased even more with the subprime mortgage crisis in 2008, Xynthia in 2010, Deep Water Horizon in 2010, Fukushima in 2011, the rail disasters of 2013, Tianjin in 2015, the collapse of the Genoa viaduct (Morandi Bridge) in 2018, etc.

It was particularly after the fatal collapse of a highway bridge in Minneapolis in 2007, that authorities and business leaders became aware that available budgets no longer made it possible to carry out grand projects, which would lead to the development of studies on structural reliability, on predictive maintenance for SSC, on ageing and durability, and on the optimized management of industrial assets throughout the lifecycle, from preliminary design to deconstruction.

1.3. Generating nuclear power

Enrico Fermi and his team at the University of Chicago were the first to develop an atomic pile. The first controlled fission reaction with electricity production was obtained on December 2, 1942. This first reactor was a fast neutron reactor.

An initial connection with an electricity network was made on December 20, 1951. This was the EBR 1 (Experimental Breeder Reactor 1) test at the Idaho National Laboratory, with an electrical production of 200 kW (for 1.4 thermal MW).

The first connections dated from the 1950s:

– Obninsk, in the USSR, on June 27, 1954 (5 electrical MW);

– Sellafield (United Kingdom), in 1956;

– Shippingsport (USA) in December 1957 (68 electrical MW);

– Marcoule G2 and Marcoule G3 (France); respectively, in April 1959 (39 electrical MW) and April 1960 (40 electrical MW).

Construction works on Chinon A1 began in February 1957 and the power plant was connected to the network in June 1963 (70 electrical MW). Nuclear sectors were then developed, especially in the 1970s, due to the oil crisis.

On December 31, 2017, worldwide there were 448 nuclear power plants (including 292 PWRs [pressurized water reactor] and 75 BWRs [boiling water reactors]) connected to an electricity network, and 59 (including 49 PWRs and 4 BWRs) were under construction. Note that 166 power plants had already been shut down. The share of nuclear power in global electricity production is 10.4%.

Since the first construction of nuclear power plants, operators and authorities have not stopped being concerned with the challenges of security and safety. Studies on nuclear safety had been systematized since the first connections. Reliability studies were developed from the end of the 1970s, leading to the publication of a book on system reliability [GON 80] and the implementation of focus groups across the nuclear industry. At the start, there were concerns above all about technical aspects, at component level as well as at system level. Since the publication of the Wash 1400 report [USN 75], which also highlighted the importance of the human factor, confirmed some years later by the TMI 2 accident (1978), the question of human reliability became an essential subject. It was in the years 1970–1980 that Probabilistic Safety Assessment (PSA) appeared, which continually progressed and developed over these years.

We distinguish three levels of PSA:

– Level 1: making it possible to analyze sequences of accidents that could lead to damage at the heart of the reactor; the core melt accident. It involves examining the response to the installation of various types of internal events (system failure, human error, etc.) or external events (earthquakes, explosions, etc.) and quantifying the probabilities.

– Level 2: uses the results of level 1, it involves evaluating the behavior of the containment and determining the nature, probability and importance of fallout from radioactive materials into the environment.

– Level 3: involves evaluating the probabilities and consequences in terms of dosimetry or contamination, and the effects on the environment and health of the population.

This work is situated mainly in the context of PSA level 1, which is the subject of international consensus.

Some years later, major accidents in the space sector (the end of the 1980s) demonstrated the impact of the organization on the behavior of equipment and installations. Finally, the possible accidental combustion of hydrogen, risks linked to natural harm, in particular to earthquakes and floods, the accidents experienced in recent years in civil engineering structures (the bridge at Minneapolis, the Morandi bridge), extreme events, have incited users to protect nuclear facilities even more, which has led to the development of structural reliability, thus encouraging reliability-based design approaches, from a design based on reliability.

Reliability should rise to many challenges:

– in first place to ensure the security/safety of nuclear facilities to protect the environment;

– to be available, to provide reliable electricity production, at the lowest possible cost;

– to optimize maintenance and support logistics to reduce their costs;

– finally to consider operational feedback from power plants now in use to better design future power plants.

Reliability has become unavoidable. In the first place, it makes it possible to understand operation feedback and to compare observations with theoretical results. Then, it makes it possible to reduce the occurrence of unfortunate situations, anomalies, faults, failures, and so to improve, using innovative or mitigation measures, the behavior of SSCs and organizational and human behaviors. To become more reliable, is to progress, is to demonstrate a qualitative and quantitative improvement in state between two dates. However, although reliability is necessary, it is not enough. It is possible to be reliable without being safe.

To be safe, it is therefore necessary to be reliable. Safety is based more on a defense-in-depth approach formed of several levels of technical and organizational protection to guarantee human safety (of the public and of operators) and the environment. The principles are the following [INS 97, LLO 06, MAG 06]:

– the prevention of failure, which requires a reliable design of the facility and a reality that conforms to it;

– monitoring of the installation, which makes it possible to detect potential defects and so to maintain a satisfactory level of reliability when in operation;

– and if, despite prevention and monitoring, a failure occurs, the limitation of its consequences because of reliable safeguarding systems.

Evaluation of the safety level requires verification of the last two principles and relies on evaluation of operational reliability. Verification of compliance with regulatory requirements (margins, barrier integrity, etc.) is the first principle: this is a deterministic evaluation. Taking account of the existence of a residual risk, this deterministic verification is completed by a more realistic probabilistic evaluation of the reliability level of the SSCs integrating uncertainties, because of the probabilistic approach.

Table 1.1, taken from Magne et al. [MAG 06], compares the positioning of probabilistic and deterministic approaches.

Deterministic	Probabilistic
Certainty	Uncertainty
Answer yes/no	Probabilistic quantification
Physical models	Physical or statistical or physical-probabilistic models
Characteristic values (upper, average, etc.)	Random variables
“Envelope” evaluations, security coefficients, margins	Realistic evaluations, important factors, hierarchy of uncertainties

Table 1.1. *Positioning of probabilistic and deterministic approaches*

The major accident examined by nuclear safety studies is the melting of the core (overheating, then melting of the fuel rods in the uncooled reactor). Some probabilistic objectives have been fixed for this major accident, as at the end of the

1930s (see section 1.2.2). For first-generation French nuclear power plants (CP0), the objective is a frequency (see Chapter 3 for the meaning of this term) of 5×10^{-5}/(reactor × years). It was then fixed at 1×10^{-5}/(reactor × years) for second generation nuclear power plants, then at 1×10^{-6}/(reactor × years) for the European Pressurized Reactor (EPR). It can also be noted that operational feedback from nuclear power plants in use worldwide reached 17,266 (reactor × years) on December 31, 2017; at this date, 448 nuclear units (including 292 PWR units) were connected to the network. This was not at all sufficient and did not allow for these objectives to be measured. It will be seen in some chapters of this book that feedback (input data) is a real problem with small data. Table 1.2 compares allocated, planned and observed performances of the different PWR platforms used in France, making the hypothesis that at any instant the frequency of core meltdown is the same. We note a small gap between what has been allocated and what has been predicted, and a more significant gap between what has been predicted and what has been observed. What has been observed consists of the following accidents: Three Mile Island 2, Chernobyl 4, Fukushima 1-2-3, which are five core meltdowns observed. It would take a sample a whole 10 times larger to verify the safety objective. The appearance of a single accident "across all sectors" multiplies the accident frequency observed by a factor in the order of 1.20.

<table>
<tr><th>Platform</th><th>Allocated objective</th><th>Calculated predicted values
(2010)</th><th>Value observed – Operational feedback from all sectors
(estimation on 31/12/2017)</th><th>Value observed – Operational PWR feedback
(estimation on 31/12/2017)</th></tr>
<tr><td>900</td><td>5×10^{-5}</td><td>7.7×10^{-6}</td><td rowspan="3">29×10^{-5}
/(reactor × years)
90% two-sided confidence interval
$[11 \times 10^{-5}; 61 \times 10^{-5}]$</td><td rowspan="3">$9 \times 10^{-5}$
/(reactor × years)
90% two-sided confidence interval
$[0.5 \times 10^{-5}; 41 \times 10^{-5}]$</td></tr>
<tr><td>1300</td><td>10^{-5}</td><td>7.2×10^{-6}</td></tr>
<tr><td>N4</td><td>10^{-5}</td><td>5.7×10^{-6}</td></tr>
</table>

Table 1.2. *Rates of major accidents involving nuclear electricity production*

Taking account of these stringent requirements, new methods of estimating reliability have appeared in recent years, from experience feedback to big data collection, we are seeing the ever greater development of reliability-based design, whether this is at the design stage or at the stage of life extension. These new methods make it possible to highlight the critical SSCs by calculating important factors, to optimize performances and lifetimes and to anticipate failure or ageing, thus leading to more reliable and safer installations and equipment.

However, reliability estimation is highly dependent on feedback, that is, on the data available, its quantity, quality and in particular its relevance. Many families of probabilistic approaches will thus appear. They are briefly cited in Table 1.3. We note that there are links between these different approaches. Thus, the rate of instantaneous failure $\lambda(t)$ defined above, the base size of the frequency reliability, is linked to the threshold crossing rate, the size used in the structural reliability depending on time to calculate the probability of cumulative failure over a given period [ARD 13].

Amount of operation feedback data	Zero or very little	Some data (<7)	More data (<30)	A lot of data
Sample	Non-existent or very small	Small	Classical, standard	Very significant
Examples	Rare events, structural reliability (passive components)	Reliability mainly in design	More reliability in use – maintenance (active components)	
Nature of data	Context data, physical data	Heterogeneous data, expertise	Data on failure and maintenance	Data on monitoring, measurable effects
Approach	Physical approach, physical reliability approach, theory of extreme values	Bayesian approach	Frequency approach, classical approach	Frequency approach, big data approach
Name	Structural reliability, extreme values analysis	Judgment/ knowledge-based probability	Statistical probability	
Methods	Combining the description and propagation of uncertainties	Combining objective and subjective data	Interpreting the objectivity of the operational feedback	
Results	Probability of failure, uncertainties, influencing factors	Laws, uncertainties, credibility	Laws, uncertainties, confidence	Short-term predictions, risk factors, analysis of degradations
Investment	High	Moderate	Moderate	High

Table 1.3. *Families of probabilistic approaches depending on the operational feedback available*

Table 1.4 summarizes the main dates in reliability in nuclear electricity production.

Date	Deed
First-century BC	*Utilitas, firmitas* (Vitruvius)
1654	The geometry of chance (Blaise Pascal)
1738	Definition of risk, decision-support (Daniel Bernoulli)
1763	Bayes' theorem
1930–1940	The first probabilistic objectives
1939	Weibull's law
1942	First production of nuclear energy
1949	FMECA
1951	First connection to an electricity network
1967	Farmer's diagram, start of the probabilistic approach
1974	Mil-Hdbk-217 B
1975	Wash 1400 (first PSA)
1978	TMI 2 accident, human reliability
1990s	RCM, LCM, AM
2010s	Development of *big data*, conditional maintenance
December 31, 2017	448 units connected to a network worldwide

Table 1.4. *Significant dates in reliability for nuclear energy production*

1.4. Presentation of the book's content

Table 1.5 lists the different chapters of this book.

Chapter 2 addresses the operation feedback on which any reliability studies rely, its concepts and its strategic role. It is held back mainly by data on failure and maintenance (failure data) and data on events. These two types of data are the most widely used in risk-control, operational safety and dependability studies. These data should be managed, analyzed qualitatively, then processed quantitatively and interpreted. Reliability experts must be familiar with the data they are using.

Chapters 3–5 focus successively on installations' systems, structures and components (SSG). For each of these chapters, industrial applications will illustrate the approaches specified.

Chapter	Author(s)	Title
1	André Lannoy	Aims and Introduction
2	André Lannoy and Emmanuel Remy	Input Data: Operational Feedback and Expertise
3	Marc Bouissou	The Principles of Calculating Reliability in Level 1 PSAs
4	Emmanuel Ardillon	Structural Reliability: General Presentation, Applications for Nuclear Power Plants
5	Emmanuel Remy	Probabilistic and Statistical Modeling for the Reliability of Industrial Equipment
6	Nicolas Dechy, Yves Dien and Jean-François Vautier	The Human and Organizational Dimensions of Reliability and Nuclear Safety
7	André Lannoy and Emmanuel Remy	From Too Little to Too Much: The Impact of Big Data
8	André Lannoy *et al.*	Conclusions and Perspectives

Table 1.5. *Contents of this book*

Chapter 3 presents the modeling techniques and general principles of reliability studies of systems in Level 1 probabilistic safety evaluations. These principles rely on the Wash 1400 report from 1975 and have never been questioned since this date. The chapter explains the theoretical foundations of calculations, presents their implementation and proposes new methods under development with the aim of further improving the relevance of estimations.

Chapter 4 is devoted to the analysis of structural reliability. Since the consequences of structural failures are generally very serious, the objectives of reliability are high. The estimations are made on the basis of the stress-resistance method. The various uncertainties (random, epistemic) are quantified, propagated in the physical–probabilistic pairing model, then hierarchized. A sensitivity analysis makes it possible to highlight the most influential factors or risk factors. The most widely used methods (Monte Carlo, FORM, SORM, etc.) and the most recent (polynomial chaos, AK method, etc.) are described.

Chapter 5 concerns the main models for components. It emphasizes the diversity of approaches, due to the nature of the equipment, to the phenomena involved, to the impact of maintenance and to its effectiveness, to the variety of industrial needs, to the quality of feedback and to its relevance, and to the complexity of the real world.

Nuclear accidents or those experienced in other industrial sectors have shown the need to consider the human and organizational dimensions to improve reliability and nuclear safety. Methods have emerged in the 1970s and have been developed since the start of the 1990s, especially for probabilistic evaluations of safety and for design studies for new reactors. Chapter 6 gives a state of the art of these reliability methods for work situations and the organizations used to improve reliability. These methods have contributed to the evolution and consolidation of good practice.

Chapter 7 projects into the near future. Data from monitoring will become more numerous and more massive, and will make it possible to determine the precise conditions in which equipment is used. The chapter aims to demonstrate the contributions of analysis and processing that the methods linked to *big data* offer (automatic language processing, machine learning, etc.) to analyze a situation, estimate reliability, predict the performance of a system or component and to optimize their predictive maintenance.

Finally, Chapter 8 will conclude the work and offer the reader some perspectives, operational or R&D, which seem to be a priority for development in the near future.

1.5. References

[ARD 13] ARDILLON E., "Time-variant reliability: some theoretical considerations for industrial practice", *Proceedings of the International Conference on Structural Safety and Reliability*, pp. 5203–5210, 2013.

[BAR 65] BARLOW R.E., PROSHAN F., *Mathematical Theory of Reliability*, John Wiley & Sons, Inc., New York, 1965.

[BAY 63] BAYES T. "An essay toward solving a problem in the doctrine of chances", *Philosophical Transactions Essay LII*, pp. 370–418, 1763.

[BAZ 66] BAZOVSKI I., *Fiabilité – Théorie et pratique de la sûreté de fonctionnement*, Dunod, Paris, 1966.

[BER 38] BERNOULLI D., *Specimen theoriae novae de mensura sortis, In Commentani Academiae Scientarum Imperialis Petropolitanae 5*, Saint-Petersbourg, 1738.

[BUF 76] BUFFON, *Histoire naturelle, générale et particulière*, servant de suite à la Théorie de la terre & d'introduction à l'histoire des minéraux, Imprimerie Royale, Supplément, Tome Troisième, voir le Onzième Mémoire : expérience sur la force des bois, pp. 158–261, 1776.

[CHA 68] CHAPOUILLE P., DE PAZZIS R., *Fiabilité des systèmes*, Masson, Paris, 1968.

[DEF 37] DE FINETTI B., *La prévision : ses lois logiques, ses sources subjectives*, Annales de l'Institut Henri Poincaré, Paris, 1937.

[EPS 53] EPSTEIN B., SOBEL M., "Life testing", *Journal of the American Statistical Association*, vol. 48, pp. 486–502, 1953.

[FAR 67] FARMER F.R., "Siting criteria – A new approach", *Proceedings of the IAEA Symposium on Nuclear Siting*, IAEA, STI/PUB/154, SM-89, pp. 303–329, 1967.

[GON 80] GONDRAN M., PAGÈS A., *Fiabilité des systèmes*, Eyrolles, Paris, 1980.

[INS 97] INSAG, "La défense en profondeur en sûreté nucléaire", Rapport du groupe consultatif international pour la sûreté nucléaire, INSAG-10, IAEA, Vienna, 1997.

[KAU 69] KAUFMANN A., *La confiance technique, théorie mathématique de la fiabilité*, Dunod, Paris, 1969.

[LAN 94] LANNOY A., PROCACCIA H., *Méthodes avancées des bases de données du retour d'expérience industriel*, Prefaces by Jean-Gérard Roussel and Yves Bertrand, Eyrolles, Paris, 1994.

[LAN 06] LANNOY A., PROCACCIA H., *Evaluation de la fiabilité prévisionnelle*, Editions Tec&Doc, Paris, 2006.

[LAN 08] LANNOY A., *Maîtrise des risques et sûreté de fonctionnement : repères historiques et méthodologiques*, Lavoisier, Paris, 2008.

[LAN 18] LANNOY A., PLANCHETTE G., *De la sûreté de fonctionnement à la maîtrise des risques : un regard sur le passé pour mieux se projeter dans l'avenir*, Actes du Congrès λμ21, Reims, available at: www.imdr.eu, 2018.

[LEM 05] LEMAIRE M., CHATEAUNEUF A., MITTEAU J.-C., *Fiabilité des structures – Couplage mécano-fiabiliste statique*, Hermès, Lavoisier, Paris, 2005.

[LEM 14] LEMAIRE M. *Mechanics and Uncertainty*, ISTE Ltd, London and John Wiley & Sons, New York, 2014.

[LIG 74] LIGERON J.-C., MARCOVICI C., *Utilisation des techniques de fiabilité en mécanique*, Technique & Documentation, Paris, 1974.

[LLO 06] LLORY M., DIEN Y., *Les systèmes sociotechniques à risques : une nécessaire distinction entre fiabilité et sécurité*, Performances, no. 30, 31, 32, 2006.

[MAG 06] MAGNE L., VASSEUR D. (eds), *et al.*, *Risques industriels – Complexité, incertitude et décision – Une approche interdisciplinaire*, Editions TEC&DOC, Lavoisier, Paris, 2006.

[MAY 26] MAYER M., *Die Sicherheit der Bauwerke und ihre Berechnung nach Grenzkräften anstatt nach zulässigen Spannungen*, Springer Verlag, Duisburg, 1926.

[PRO 92] PROCACCIA H., PIEPSZOWNIK L., CLAROTTI C.A., *Fiabilité des équipements et analyse statistique décisionnelle fréquentielle et Bayésienne*, Collection de la direction des études et recherches d'Electricité de France, Eyrolles, Paris, 1992.

[REA 97] REASON J., *Managing the Risks of Organizational Accidents*, Ashgate Publishing Ltd., Farnham, UK, and Burlington, USA, 1997.

[SCH 69] SCHWOB M., PEYRACHE G., *Traité de fiabilité*, Masson, Paris, 1969.

[SWA 83] SWAIN A.D., GUTTMANN H.E., *Handbook of Human Reliability Analysis with Emphasis on Nuclear Power Plant Applications*, NUREG/CR-1278, Washington, D.C., US Nuclear Regulatory Commission, 1983.

[TAL 10] TALEB N.N., *The Black Swan – The Impact of Highly Improbable*, 2nd edition, The Random House Publishing Group, Penguin Books, London, 2010.

[USN 75] US NRC, "Reactor safety study: an assessment of accident risks in US commercial nuclear power plants", WASH-1400, NUREG675/014, 1975.

[VIL 88] VILLEMEUR A., *Sûreté de fonctionnement des systèmes industriels*, Collection de la direction des études et recherches d'Electricité de France, Eyrolles, Paris, 1988.

[WEI 39] WEIBULL W., "A statistical theory of strengths of materials", Ing. Vetenkaps Akad. Handl. No. 151; "The phenomena of rupture in solids", Ing. Vetenkaps Akad. Handl. No. 153, Royal Swedish Academy of Sciences, Stockholm, 1939.

2

Input Data: Operation Feedback and Expertise

Reliability estimation requires data, a great deal of data, including varied data on failures or events, physical data or expertise. This chapter tackles operation feedback, on which all and any reliability studies rely, its concepts and its strategic role. We will spend time especially on failure data and on event data, which are the most used in studies on risk or dependability, and we will show that any operation feedback must be analyzed qualitatively at the outset so as to provide accurate and relevant data. Finally, the reliability expert is obliged to be familiar not only with data and reliability, but also with the context of use, the operation and the comprehension of physical phenomena and human behavior.

2.1. The purposes of operation feedback

The purpose of operation feedback is to collect, archive, validate, process and analyze information specific to the behavior of installations and important equipment. Operation feedback consisted of all the technical, organizational and human facts, such as events, accidents, incidents, failures, degradations or maintenance operations, etc., observed through the operation of installations and the equipment of which they are formed, from the preliminary designs to their decommissioning. It should make it possible, at a local or centralized level, to establish the plant's health, to identify weak points (malfunctions, failure and degradation modes, causes) and, if necessary, to identify ways to improve the reliability of equipment. Installations and equipment, which it is important to monitor, play a role in the challenges of ensuring safety and dependability, protecting the environment, availability, the costs of maintenance, and dosimetry.

Chapter written by André LANNOY and Emmanuel REMY.

Operation feedback presents important challenges for a nuclear power generation company.

Managing safety and protecting environment: Operation feedback is a vital tool to know the performances of equipment important to safety. It should make it possible to establish reviews of this equipment's condition, thus making it easier for sites to self-evaluate, to identify precursor events and to obtain the quantitative data needed for probabilistic safety assessments (PSAs; see Chapter 3) and for periodically revising them and analyzing the potential impact on the environment. Monitoring equipment behavior at the local level makes it possible to guarantee the installation's safety level, and monitoring the reliability of equipment makes it possible, via PSAs, to evaluate the average safety level of installations with the same design.

Managing availability: Operation feedback on equipment that can entail a lack of availability makes it possible to establish diagnosis and there too to deduce from these, if necessary, corrective actions to ensure availability, whether at local or national level.

Managing maintenance costs: It is necessary to collect raw data on failures, degradations and maintenance operations to establish technical and economic reports on the behavior of equipment monitored locally or nationally: it makes it possible to understand deteriorations, to anticipate problems to come and to optimize preventive maintenance policies. On this last point, operation feedback is vital for diverse maintenance analyses (reliability centered maintenance (RCM)). The challenge of maintenance is probably the most motivating for an operator; in practice, it contributes a great deal to the quality of the data collected.

Managing durability and extending installations' operation: It is essential to have complete histories of SSC (systems–structures–components) if we wish to be able to appreciate the durability or the usage factor of equipment and so to contribute to this major challenge.

Helping design of future installations and equipment: By high-lighting weak spots and "diseases" in equipment, it will be possible in the future to mitigate them and to improve their reliability by ensuring that the mistakes of the past are not repeated. A design approach cannot work effectively without considering the lessons of feedback from past operations.

The main purpose of operation feedback is therefore to progress in all areas of activity (quality, safety, availability, reliability, durability, security, maintenance, costs, human and organizational factors, environment, radio-protection, communication). Operation feedback is not an end in itself. It is a means of progress, contributing to the quality (which is essentially aimed at the conformity of the installation and its equipment) and to the reliability (reliability is the extension of quality over time) of a product, a system, a process and a method of work. The feedback may be positive or negative. It is acquired from real-life situations, practices and malfunctions, whether technical or organizational.

2.2. What is operation feedback?

We can find many definitions in the technical documentation, depending on users' needs. We can point out current efforts in the context of standardization to define this concept.

We propose here some definitions that relate more to the processing of reliability.

In the first place, the dictionary definition of risk analysis and risk management [DES 05]: *formalisation de la prise en compte de l'expérience tirée d'une activité passée et de la transmission du savoir afin d'améliorer la qualité et l'efficacité des produits* (formalizing considerations developed from experience of past activity and from the transmission of knowledge to improve product quality and effectiveness), the aims being to decrease the number and gravity of errors, to redirect those processes that work (the opportunities), to improve working methods and to decrease risk and deviation from specified objectives (technical performances, quality, costs and delays). The dictionary gives sources of operation feedback: expert opinions; technical reports, actions and recommendations from reviews; analyses of failures and failures to compliance; end of project reports; customer complaints and warnings.

This is *management of technical facts and technical performances, observed throughout the lifetime, from design to withdrawal from service*; operation feedback includes three main stages: (i) collection and memorization of feedback data, (ii) validation and analysis of the information collected, (iii) diffusion and application of lessons learnt from analysis of operation feedback [LAN 94]. We note that this definition essentially concerns technical feedback.

It is *the set of methods for collecting, storing, managing and using information* to correct, adapt or improve systems currently in use or future systems; one of its objectives is to learn better from lessons from past feedback to engage in targeted actions (definition by Jacques Valancogne [LAN 11]).

Different bases for operation feedback are used in the nuclear industry:

– Operation feedback on events, containing information on accidents, incidents, etc. It is much used for operation, its impacts on safety and availability, the search for weak signals and human and organizational factors.

– Operation feedback on equipment relates to the failures and degradations observed, maintenance actions, etc. It is much used for its impacts on safety, reliability, maintenance, the costs of maintenance, durability and design support.

– Data from monitoring, now called HUMS (Health and Usage Monitoring Systems [IMD 17]), resulting from conditional maintenance. These data are undergoing very significant development at the present time due to automatic collection and the arrival of big data tools able to process and interpret it; this point will be addressed in Chapter 7; it is much used for monitoring, optimizing maintenance, durability and protecting the environment.

– Data from inspection – tests corresponding to physical defect measurement (effects that are measurable due to a physical degradation mechanism) made at more or less regular intervals of time, making it possible to estimate degradation kinetics.

– Operating statistics describe installations' usage or operation profiles (number of shutdowns, duration, availability coefficient, etc.), which are indispensable to any reliability calculation. HUMS data will also make it possible to get to know usage profiles better.

– Process data are data related to the physical–chemical operating conditions of a system. These data are very varied (temperature, pressure, chemistry, constructor, material, equipment, environment, state of the unit, etc.) and can intervene as explanatory factors for technical events. It is even more important as big data approaches and machine learning algorithms develop and aim to identify correlations between parameters (see Chapter 7). Data on materials, on loading and stresses and on the defects that potentially exist, are important to estimating a structure's reliability (see Chapter 4).

The reliability specialist will essentially use the first two types of operation feedback, based on events and on equipment, as well as expertise [LAN 01].

In fact expertise, which may be qualitative or quantitative, is a form of operation feedback that is very useful in reliability problems. It is a source of subjective information, representative of an authorized and recognized opinion, based on knowledge, training, practice and experience of experts from a particular domain, at a given instant. It makes it possible to complete, and to improve on objective data when it exists, and when it is scant, sparse, doubtful, or inapplicable, or to supplement it when data are lacking (e.g. in the case of poor operation feedback or a future problem or an innovation, etc.). It may be the only source of information available that allows predictions to be made. It makes it possible to interpret technical or organizational reliability analyses with a view to facilitate the work of a decision-maker.

In their analysis, the reliability specialist user will very often need other operation feedback data taken from databases or documents or plans, data that may involve:

– system architectures, SSC characteristics;

– design data, data on equipment (see above);

– preventive maintenance programs, data from logistical support;

– context data (environment, geography, meteorology, etc.); these data are increasingly more numerous, especially when considering extreme natural harms (earthquakes, floods, etc.) where it makes sense at the outset to define the hazard (its frequency, its more serious characteristics, the loading created);

– procedures, operating instructions, safety regulations;

– descriptions of human behavior;

– data from reliability tests or reliability handbooks;

– situations observed abroad or among competitors; these data are particularly important to safety (preventing any accidents that may happen), for detecting ageing (observing the behavior of older systems) and for performance benchmarking.

2.3. The operation feedback approach

Figure 2.1 shows the full process of analyzing a database of industrial feedback, from design to operation-maintenance, local level 1 analysis when it is carried out by sites where the technical facts are observed or after initial expertise, and level 2 analysis by an engineering team, during which incidents are analyzed in detail for preventive purposes or when qualitative or quantitative results are used in large-scale projects.

One of the critical points in analyzing operation feedback is data quality. Quality is vital, because it has a direct impact on the reliability of the results and on the interpretation of operation feedback. The issue of quality first comes into play at the collection stage, on site, better for capturing, interpreting, verifying, correcting and validating the data involved, but it appears also at the analysis stage. It consists of monitoring the accuracy (which plays a role at collection) and the relevance (important before analysis) of the data for the problem posed and explaining data that may be extreme or unusual (see section 2.5.4).

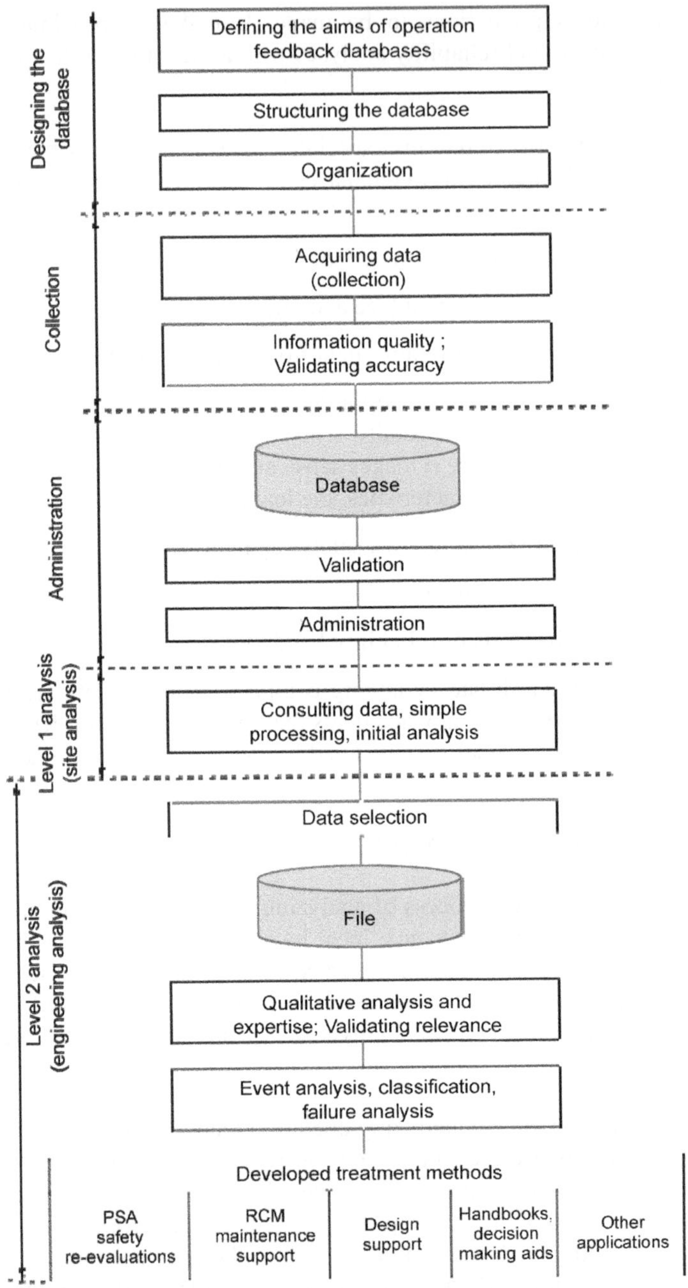

Figure 2.1. *Analysis process of an operation feedback data bank. For a color version of this figure, see www.iste.co.uk/lannoy/nuclear.zip*

A deviation is detected by the operating sites and this initiates collection of operating information and information on the situation. The site at which a piece of information originates owns this information and guarantees its accuracy and exhaustivity. Generally, the site uses its own data for its statistics, for local safety studies (calculating PSA data, etc.) or maintenance (optimizing maintenance using reliability, etc.). The site sends its data to a central administration service (administrator of operation feedback databases), which gathers all the data from different sites. This is then used by a central engineering department for the purposes of safety, optimization of preventive maintenance, performance or designing new installations or systems.

The first level at which the situation is analyzed is therefore local. The site where a situation has been detected seeks the direct or root causes as well as solutions, whether palliative or corrective. The structure of operation feedback allows it to be used to research similar situations that may be encountered in identical or like conditions, on other sites and associated intervention reports.

A second level at which the situation is analyzed in a central engineering department allows it to check the roots of the causes behind the situation, to consider potential consequences that may be generic across all installations. These analyses can potentially be communicated to all operating sites, the designer, the constructor and the regulatory authorities. They make it possible moreover to complete operating rules, maintenance policies, preventive maintenance programs, monitoring, and to carry out annual diagnosis on the behavior of equipment (reviews). Reliability studies are generally carried out at this engineering level.

Operation feedback is consequently at the root of reliability and maintenance analyses and evolutions affecting methods and SSCs: technological progress, modifications to design, maintenance optimization and ageing management. Indeed, throughout the lifecycle, it is necessary to:

– clarify operational modes, often complex;

– improve safety performances;

– adapt the frequency of maintenance operations depending on feedback;

– arrange operation and maintenance depending on technical and technological evolutions.

The database is a vital tool for operation feedback. It is useful:

– before creating a database, to define its aims well; a needs analysis is essential and it cannot be carried out without potential users of the database; this stage serves to structure the database and prepare the collection of data at sites; with the arrival of big data, we can recommend that the designer of an operation feedback database build and develop interoperable databases making it possible to gather and compare information easily with data from other, different data bases. This is now one of the aims of data lakes which are a fast storage method (see chapter 7) to avoid databases becoming "siloed";

– to organize collection on site, the quality of data and therefore the quality of the analyses that will be made from it is very dependent on its collection;

– to have a user-friendly software tool available; online help, fast response times from the software and easy access to archived data are among the factors that play a role in good data quality.

Below, we will focus on "event" operation feedback, aiming to improve safety and availability performances, then on "equipment" feedback, very similar to the technical needs of reliability. Since the 1960s, we have known that reliability is strongly influenced by the effectiveness of maintenance, with maintenance leading to the rejuvenation of equipment if it is done well, and playing a disruptive role in estimating reliability [BAR 65] (on this subject, see section 5.4.4 of Chapter 5).

2.4. "Event" operation feedback

Event feedback contains the story of the situations observed. Mainly, this is all the events relating to availability or safety:

– shutdowns or scrams of the plant due to an incident internal or external to the power plant;

– all the equipment failures observed during operation or after shutdown;

– all events considered significant from a safety perspective, depending on the criteria chosen by the safety authorities.

The situation is described in the event database from the occurrence of the first root cause in a normal situation to return to a normal, operational situation, after corrective action [ESR 09]. The data collected concerns the following:

– *Identification and description of the event*: the circumstances and modes by which it is detected, the state of the plant at the moment of the event and the characteristics of the event.

– *Causal analysis of the event, its technical causes, causes linked to organizational and human factors*: technical analysis of the event, the development of a cause tree from direct causes to root causes (or deeper causes) and their associated elements; the grouping of elements of events into a family; characteristic criteria relating to the carrying out of activity by actors, the factors (environmental, sociological, etc.) that have had an influence on how the activity is carried out, criteria relating to the organization of activities, analysis of defects or difficulties in the instructions or in procedures, in user assistance and in the man/installation interface [GRO 17].

– *Description of the real and potential consequences*: the identification and description of real consequences in terms of operating performances (availability, safety) and in terms of failures or degradations of equipment; technical, human, organizational, economic, financial and environmental impacts; analysis of the potential consequences (what would have happened if…?), this last point is very important and extremely useful for anticipating potential malfunctions (such as weak signals) and their potentially very grave consequences.

– *Analysis of corrective actions*: corrective actions made at the local level, in terms of equipment or organizational arrangement, are generally made in emergencies: re-establishing functioning as quickly as possible, managing the crisis; these analyses are generally immediate and qualitative; preventive measures are on the contrary decided at local or central level and are generally deferred; they are the subject of qualitative and quantitative analyses, a statistical processing, and pedagogical sharing of lessons from operation feedback to other installations of the same type.

– *Specific analyses*, useful at national level for designing new installations or for managing lifespan; they mainly involve the domain of safety, maintenance, human and organizational factors, crisis management, equipment and materials.

Once the event analysis has been carried out, it makes sense to publicize it, develop it and share the lessons that can be learnt from it [ESR 09, GRO 17].

2.5. "Equipment" operation feedback

Maintenance models and collection problems are mentioned in the following sections, taken from [LAN 11] and also updated.

2.5.1. *The maintenance model: an approach according to function*

The process function (e.g. providing a source of cooling) is a function that must be achieved by a set of equipment within a system.

Different elementary technological functions (ETF) play a role in process functions (e.g. setting a fluid in motion, exchanging heat). With each ETF thus defined, there is an associated piece of equipment whose function is to carry it out. Thus, to set a fluid in motion, we can use either an electric motor pump, or a turbopump in the case of a liquid, a fan motor in the case of a gas.

The base technological function (BTF) clearly indicates the choice of equipment. Thus, to provide a rotational force for the pump, we can use an electric motor, a turbine, or indeed a diesel motor. It is therefore sufficient to define the actions expected of this equipment, for example, transforming electrical energy into a rotational movement.

The base technological sub-function is of particular interest, because it is this that allows the precise choice of sets of components to be examined. In the case of a pump, the sub-function "seal" will mean that a mechanical seal should be used, or even a simple packing box. It is at this level that maintenance operations are generally carried out.

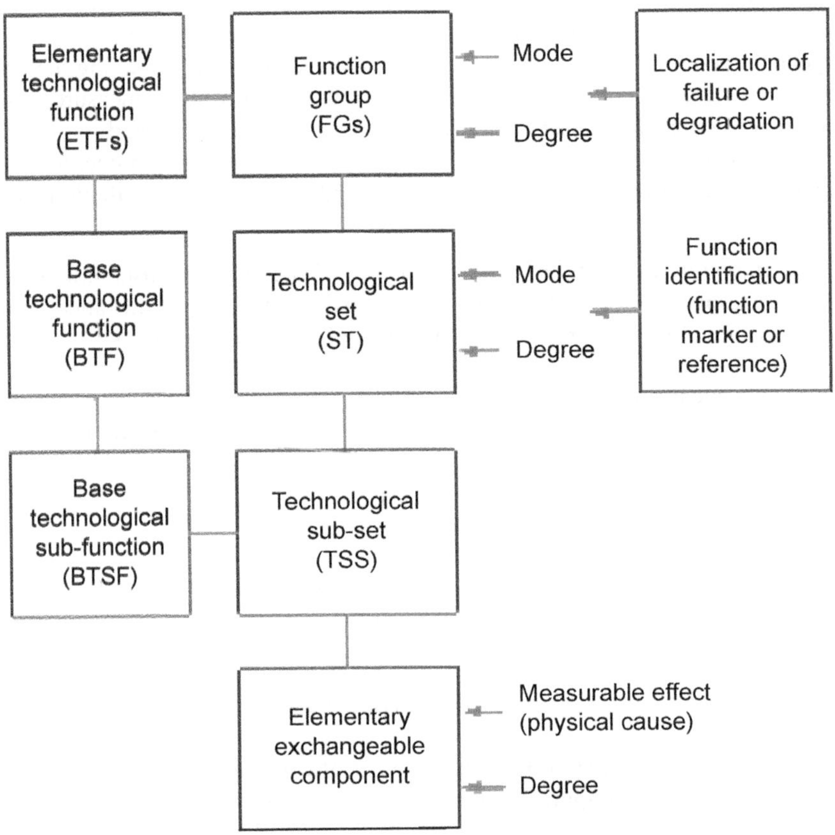

Figure 2.2. *Function-equipment breakdown for a piece of equipment. For a color version of this figure, see www.iste.co.uk/lannoy/nuclear.zip*

Failure mode (effect through which the failure is observed), which is an important parameter for reliability and safety studies, is defined at the level of the ETF and the BTF (e.g. untimely Sudden or untimely stop and leak). At the level of basic components or spare parts (the lowest level of the function-equipment cause tree), it is necessary to know the cause of the failure. A list of measurable effects or physical causes or degradation mechanisms should be established with the help of equipment experts. The component is the spare part, the final leaf on the function-equipment cause tree, which is replaced during maintenance action.

This breakdown is finally a maintenance model, serving to structure the data. The breakdown is generic and carried out using FMEA (Analysis of Failure Mode & Effects Analysis) made for each piece of equipment monitored. Use of this breakdown ensures not only consistency between safety and maintenance studies, but also coherence with design studies. In particular, the components modeled in probabilistic safety evaluations (PSAs, see Chapter 3) can be associated, without ambiguity, with elements of the equipment tree used as far as the lowest level by maintenance. Figure 2.2 shows the links between functions and equipment. ISO 14224 standard: 2016 specifies the same approach.

2.5.2. *Failure analysis*

Failure analysis consists of identifying, up to the finest function-equipment level, the causes, degradation mechanisms, mode, potential and real effects of a failure. It is on this information that preventive or corrective measures will be based, which will be retained to re-establish function or to avoid a new failure. The analyst can investigate, carry out tests or physical calculations, in short use any method deemed necessary to determine the causes of the failure. Use of this equipment tree defines, without ambiguity, the limits of the equipment. Logical analysis of the failure consists therefore of monitoring the causal tree from the level impacted by the failure to the finest unit affected. The approach is descending, that is, deductive, from the failure observed on the equipment to the affected component, to the mode and the measurable effect. Adoption of this approach facilitates the work of whoever writes the failure report.

Figure 2.3 defines the basic data to be collected, which is important to safety studies (left-hand column) and for maintenance studies (right-hand column).

The equipment to be monitored is the equipment recognized as critical. It meets the following criteria:

– critical for safety;

– responsible for production losses, critical for availability;

– with high maintenance costs;

– critical for dosimetry.

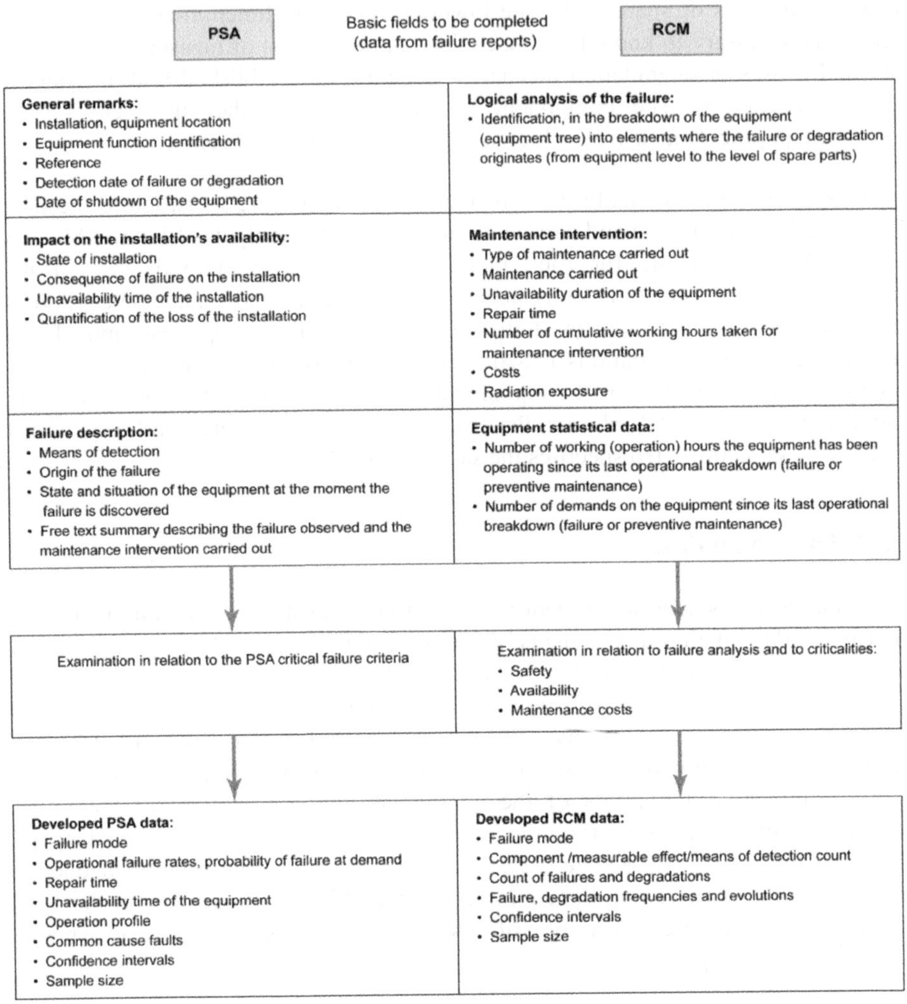

Figure 2.3. *Comparison of PSA (safety) and RCM (maintenance) analyses – The necessary data. For a color version of this figure, see www.iste.co.uk/lannoy/nuclear.zip*

Any maintenance operation on a piece of equipment to be monitored must be recorded in the operation feedback database:

– the failure and associated corrective maintenance action;

– degradation detected during a preventive maintenance operation;

– “visiting” maintenance operation or inspection.

This information makes it possible to reconstruct the full life history of the equipment. It can be noted that in computer-assisted maintenance management (CAMM) databases are linked to “equipment” databases. They are necessary but not however sufficient alone, as they do not generally provide information linked to dependability.

2.5.3. *Failure criteria*

Failure in the sense of maintenance is a complete failure, with loss of the equipment’s effective functioning. A complete failure is a critical failure in terms of safety if, in addition, specific safety criteria are not met. Safety analysis means verifying that safety checks can be carried out at least once. The duration of the check is important. If the check cannot be carried out, then we witness a critical failure in terms of safety. For example, a failure considered complete by maintenance (replacement of equipment) can be considered non-critical for safety if the equipment can assure its mission at least once. Similarly, a simple degradation for maintenance (for example, wear) is a critical failure for safety if this degradation can lead to the loss of function while the check is being carried out, if the equipment is needed. The degree of failure (failure or degradation) is defined using general operational rules.

The summary of the failure is used to justify the analysis. In addition, one can consider tracing criteria relating to operating rules or to safety, making it possible to assess the severity of the anomaly (diagnosis information). This traceability can be useful in cases where information from the site is validated *a posteriori* or where data are re-examined. Note that current natural language processing tools can facilitate this analysis [IMD 13].

2.5.4. *Data quality*

In the field of operation feedback, perhaps more than in other fields, data quality is vital, since it has a direct impact on the reliability of results and the interpretation of operation feedback. With data that have little credibility, results too have little credibility and the database inevitably disappears [LAN 94, PET 99].

Obtaining good-quality, validated data requires implementing processes that will determine the accuracy and relevance of operation data, for a specific objective. Carried out at analysis level, this process makes it possible to obtain, from raw data, raw data that are validated for this aim. The quality of operation feedback data matters at two levels.

2.5.4.1. *At the level of information collection*

The collection task is difficult, since the officer, an expert technician, who is responsible (the report-writer) absolutely must interrogate, analyze and appraise the information before it is introduced into the database. Verifiers should validate the data collected by the report-writer, complete it and enrich it. Their contribution will be focused more on the difficult data needed for risk analysis: failure mode, degradation mechanism, specific impacts and analyses concerning safety, maintenance, human and organizational factors, barriers and their effectiveness. We would like to point out that a cross-validation (e.g. redaction by an operation officer then verification by a maintenance officer, or vice versa) seems to be the most effective form of validation. Furthermore, lack of immediate profitability, incomprehension by management of this onerous task, the lack of motivation on the part of representatives responsible for collection, the overly systematic replacement of these operators and the absence of immediate operation feedback, are a number of factors that can harm the quality of the database content.

Many elements of a record (e.g. in the case of an event record: the description, the circumstances, the causes) require expertise on the part of the report-writer, expertise that calls on a good knowledge of how the installation and their equipment are operated.

The quality of operation feedback data is already at play at collection level, on site, where it is easier to pick up, interpret, verify, correct and validate the data involved in it. Indeed, only the site where the operation feedback data originates, the owner of its own data, is able to correct and validate it.

2.5.4.2. *At analysis level*

This is the most critical point of operation feedback analysis, since it is fundamental. It is on the quality of the data that the credibility of the results and their interpretation depends. Data, models and results should always be validated by the user before any processing (see also Chapters 5 and 7).

Reliability specialists should first select, intelligently, the raw data relating to their objective and problem. Quality at analysis level consists of monitoring the accuracy and relevance of this data sample for the problem posed. Accuracy and relevance seem to us to be the two essential qualities. Quality has many other characteristics: sustainability, sensitivity, completeness or exhaustiveness, which are judged with the help of qualitative approaches. The reliability experts responsible for validation during analysis, then for the interpretation of the results, use their knowledge of the installation, equipment and operation. They must therefore have knowledge other than knowledge of the data, on reliability, on the installation and its equipment, on the physics of phenomena and on human behavior. Analysis of failure is long, fastidious and thankless, since it requires "visual" or "manual" analysis of a large number of files, but it is also the most rewarding phase, as this phase alone makes it possible to improve equipment and installations, which remains the main aim of operation feedback. Consequently, reliability specialists must validate the data they use. Their aim is to reduce uncertainty and to create confidence in its results. We must emphasize the caution they must exercise on passing information from data to reliability results.

2.6. Reliability analysis

In the previous sections, we have seen that any analysis of operation feedback requires a "visual" analysis of the information quality and its relevance for the problem posed. The expert analyst should display rigor, good sense, an ability to bring ideas together and to synthesize and should quite clearly know the data, operation, equipment and reliability. Any quantitative analysis requires at the outset a qualitative analysis, a validation of the data, then a quantitative processing for reliability estimation. Interpretation requires a return to raw data, to knowledge of operation, of usage and the equipment's behavior.

We saw in Chapter 1 (Table 1.3) that the reliability estimate depends greatly on the size of the operation feedback sample. The estimate depends also:

– on the nature of the components studied;

– on characteristics linked to the data and their history.

2.6.1. *The components studied*

We can distinguish several categories.

a) *Non repairable active components* (joints, seals, etc.) have an individual lifespan; this is very often a spare part or the terminal leaf of a device's function-equipment tree. The component undergoes multiple degradation mechanisms. When the component fails or is at the end of its life, it is replaced by a new component in the context of corrective maintenance if it is failing, or in the context of preventive maintenance if the maintenance engineer estimates that it may fail before the next maintenance intervention. For the reliability estimate, we seek in particular in addition to operating times, the failure mode, the degradation mechanism, the number of operational failures during the period studied and the number of failures at demand.

b) *Repairable active components* (pumps, taps, etc.) are subject to corrective or preventive maintenance operations that are supposed to maintain them in a good operational state or even to rejuvenate them, and so postpone failure if this maintenance is done well. Preventive maintenance has the aim of reducing the probability of failure. In the same way as mentioned before, we seek the number of situations per time interval for a population, operating times and the succession of corrective and preventive maintenance events.

c) *Active components on standby* are components that are not operational, but which should ensure safety functions when they are required. For these components, access to the number of demand failures (start-stop) is an important condition.

The points (a)–(c) will be addressed in Chapter 5.

d) *Passive components that are not repairable or difficult to repair* (e.g. pipe networks, or a reactor vessel considered to be non-maintainable), experience very few failures, or indeed none. There are few failure modes and few degradation mechanisms (corrosion, thermal fatigue, etc.). Reliability methods necessitate building a probabilistic model from a deterministic physical model and especially introducing distributions of more physical input data (such as geometry, existing faults and stresses) (see Chapter 4). It is often possible to have quantitative measures of measurable effects (depth of crack, length of crack, loss of thickness, etc.) and therefore to determine a degradation kinetic that can be compared to a threshold limit conveying either end of life or the remaining life (the Remaining Useful Life (RUL)). We also note the fatigue monitoring of important components for safety (e.g. the expansion line of the pressurizer) and subject to temperature–pressure cycles, so to thermo-mechanical stresses. The different cycles observed in production are counted. This is the accounting of situations. A usage factor can be determined in real time, which thus makes it possible to determine a remaining lifespan depending on operation predictions. The arrival of big data and the collection of monitoring data should have a great impact on this application and favor monitoring the performances of other passive components.

2.6.2. *Data characteristics*

The essential characteristic of technical operation feedback is that it is found in a very censored environment (a device's date of censor is a date from which a device that has still not failed is no longer observed for various reasons), sometimes, or indeed often, aggravated by a small sample size. It therefore makes sense always to determine the uncertainties and intervals of confidence or credibility.

Data are collected on a population of identical equipment, in the same conditions of design, construction, environment, operation and maintenance. It is always necessary to estimate the size of the available sample from "equipment" databases or databases of operating statistics. The operating and maintenance conditions should be representative of future conditions of use.

Five types of data can be distinguished:

a) *Complete data*: They correspond to instants of failure and to corrective maintenance operations to re-establish function.

However, not all devices are faulty. One characteristic of survival data is the existence of incomplete data, called censored data.

b) *Type I right censored data*: Observation is stopped at a pre-determined instant and therefore, data on failure and right-censors are counted. These cases of right-censored data are the most common in industrial operation feedback. The equipment has not experienced failure during the last observation (incomplete data). When there is right-censoring, not all lifetimes are observed. What is known, however, is that they are above a certain value.

c) *Type II right censored data*: The components are observed and the failures counted, up to a number fixed at the outset. At this instant, the equipment has already experienced failure (complete data). Other equipment has not experienced failure: this is the data known as right-censored, corresponding therefore to periods of correct functioning during the observation period. Right-censors are therefore, either periods of correct functioning (e.g. when the component is found to be neither faulty nor degraded after observation) or preventive maintenance time (when the component can of course be degraded, but where its function is assured, even if it is altered).

d) *Left-censored data* correspond to data on components that are known to have broken down before the instant of observation, but whose data of failure is now known exactly. This moment of failure pre-dates the moment of observation.

e) *Interval-censored data* correspond to an operating time interval often fixed in advance, between a starting moment (right censor) and a final moment of observation (left censor), during which the failures are observed. The only information known is what happened between the two instants. It is a relatively common case since equipment can be monitored intermittently (especially for maintenance; the monitoring will generally be continuous if the equipment is important to safety). These censors by interval are mostly found when verifying the effectiveness of a maintenance action or the effectiveness of design modification or we attempt to detect ageing.

Survival times are often obtained partially because of this censoring process. These incomplete data (cases b–e) are called censored. Although it is less informative than complete data, censored data provide non-negligible information on the reliability of equipment: they are therefore indispensable to analysis and should be handled using adapted methods (see section 5.5 of Chapter 5).

2.6.3. *Principles of simple reliability data estimation for PSAs*

A key element is the selection of the sample: what are the representative devices? Over what period are they studied? Is it necessary to enrich the sample? How big is it?, etc.

Estimating reliability requires assembling data other than failure times, especially the impacted element of the equipment, the failure mode, operating statistics and the summary for aiding interpretation.

For each failure mode, there is a linked reliability indicator, for example, a failure rate. We give some examples of simple calculations, corresponding to the exponential law conveying the lifetimes of a device whose random breakdowns survive with a semi-constant failure rate λ (see also Chapter 5; [LAN 94, MAR 74]). To make this estimation, the reliability specialist should determine the failure's criticality for safety at the outset, from general operating rules and from operation feedback.

Frequentist estimation will generally be used as soon as the number of failures observed is more than 6. A confidence interval can be estimated, as well as the error factor. These parameters are used to evaluate the influence of a device's reliability parameters on a system's reliability, in other words to evaluate the robustness of the system and the importance of the device, all other things being equal.

Bayesian estimation is used when the data are low in number and when there is *prior* information in the form of a probability distribution for the failure rate to be estimated (see chapter 5). The posterior distribution of failure rate may therefore be the average of the *posterior* distribution. We can then calculate a credibility interval and an error factor. Bayesian estimation is always used when we wish to update old data or individualize the reliability data for a site.

In the case of zero failure, Bayesian methods (Chapter 5) or stress-resistance analysis methods (Chapter 4) are more appropriate. However, regulations for nuclear and oil and gas sectors specify the formula λ = 0.7/cumulative operating time, corresponding to a quantile of 50%, the average being assimilated to the median.

– Frequentist estimation (from raw observations from operation feedback):

- Operation failures:

λ = Number of critical failures/cumulative operating time.

- Failures on demand:

γ = Number of critical failures/cumulative number of demands.

– Bayesian estimation (use of *prior* information on the failure rate):

- $P_{posterior}$ (λ/observations) ~ P(observations/λ) × P_{prior} (λ);

- Estimator λ = $E_{posterior}$ (λ/observations).

The manner in which these formulas are obtained is detailed in Chapter 5. We recall that these formulas rely on the strong hypothesis of a failure rate that is constant over time.

A reliability data handbook provides the reliability parameters, which are data estimated from validated operation feedback and used in studies on safety in particular in probabilistic safety evaluations, risk analyses and the design phase. Creating such a handbook is a long and difficult task, given the need to collect and "manually" validate the useful information set before any processing. These handbooks are also useful for monitoring quality and the production process. Generally, these handbooks relate to safety, and consider only those failures and degradations that are critical for safety, data on maintenance being considered to be part of industrial property. On the contrary, "safety" data involve only a more restricted number of devices, those considered important to safety. These safety data conveys transparency vis-à-vis safety, which can only help brand image. These handbooks are therefore useful for providing a predictive reliability estimate in the design phase, as the electronics and electrotechnical industries currently do. This provisional estimate can also serve as prior information in a Bayesian approach (Chapter 5). Data thus published should be managed, with caution, and after adaptation to conditions of environment, operation or maintenance. We can list some handbooks used in the nuclear sector: EG&G (1990, Idaho National Engineering Laboratory), EIReDA'2000 (2000, European data on the nuclear sector), NPRD-95 (1995, Non electronic Parts Reliability Data), OREDA (2013, 6th edition, Exploration and offshore production), T-Book (2005, 6th edition, Reliability data of components in Nordic nuclear power plants) and TecDoc 508 (1989, IAEA, Survey of ranges of components reliability data for use in probabilistic assessment).

2.7. Conclusion

Operation feedback is simultaneously:

– a historic archiving tool for operation data;

– a contributing tool for managing the challenges of safety and maintenance costs;

– a design support tool;

– a tool for improving skills;

– a tool for managing the lifecycle of installations and their equipment, especially for managing ageing;

– a communication and exchange tool within local and central engineering departments at industrial sites.

Analyzing it makes it possible to detect problems, analyze them, trace them, take corrective measures, verify the effectiveness of these measures and include them in our "technical heritage" (positive operation feedback).

Operation feedback is integrated therefore quite naturally into the information and management systems of industrial sites. Although it requires a substantial financial investment in the long-term, it is a source of profit and a prospective tool.

Operation feedback is the basic material for reliability studies. In this sense, reliability specialists cannot be content with using data for processing. They must select a sample intelligently, prepare it, validate it, process it and finally interpret its results. Their skills are not therefore limited to reliability, and they should also have knowledge of the functioning of the installations studied and their equipment. They should interpret then explain its results and to do this work with the physics of phenomena and knowledge of human behavior in work situations.

This latter conclusion, already verified for classic reliability studies, is even truer for the use of data in a big data approach [IMD 20].

This chapter has presented the concepts of operation feedback, its methods as much for collection as for analysis, data validation and the use of operation feedback in the field of reliability. These principles remain. However, the processing of increasing quantities of data, especially with the arrival of HUMS monitoring data, is becoming a significant challenge. This is the shift from "too little to too much". All these data represent important knowledge for operation, safety, maintenance optimization, etc., knowledge that must be extracted, but which can no longer be extracted with conventional methods. It means understanding, validating, processing and understanding from data. Evaluating the quality of the data and their preparation remains a concerning problem. Analysis and interpretation depend even more than before on the quality of the data collected (see Chapter 7). In a project, the evaluation of data quality, considering the amount of data, is a challenge. The care taken in collecting the data and in its quality will make it possible to guarantee representative and robust results. The reliability specialist will then have powerful statistical tools that allow them to highlight critical points, influencing factors, correlations, etc. Data will disrupt models and their evaluation, probably at the expense of existing models. The expert will have to confirm observations of algorithms and take appropriate decisions. All these aspects will be addressed in Chapter 7.

2.8. References

[BAR 65] BARLOW R.E., PROSHAN, F., *Mathematical Theory of Reliability*, John Wiley & Sons, Hoboken, 1965.

[CAN 91] CANNON G., BENDELL A., *Reliability Data Banks*, Elsevier Applied Science, Amsterdam, 1991.

[COO 96] COOKE R.M. "Design of reliability data bases", *Reliability Engineering and System Safety*, vol. 51, no. 2, 1996.

[DES 05] DESROCHES A., LEROY A., QUARANTA J.F., VALLÉE F., *Dictionnaire d'analyse et de gestion des risques*, Hermès/Lavoisier, Paris, 2005.

[EIR 00] EIREDA 2000, *European Industry Reliability Data Bank 2000*, Crete University Press, Heraklion, Greece, 2000.

[ESR 09] ESREDA, "Guidelines for safety investigations of accidents", Guidelines, ESReDA Working Group on Accident Investigation, 2009.

[HOO 96] HOOKSTAD P., SANDTORV N.A., THOMPSON, D.W., "Practical experiences with a data collection project: the OREDA project", *Reliability Engineering and System Safety*, vol. 51 , no. 2, pp. 159–168, 1996.

[GRO 17] GROUPE DE TRAVAIL DE L'ICSI, Processus d'enquête suite aux accidents technologiques majeurs : vision d'ensemble et pistes d'amélioration, Numéro 2017-03, *Les Cahiers de la sécurité industrielle*, ICSI, Toulouse, available at: http://www.icsi-eu.org/, 2017.

[IMD 13] IMDR, "Méthodes d'analyse textuelle pour l'interprétation des REX humains, organisationnels et techniques", Document, IMdR, 2013.

[IMD 17] IMDR, "Health and usage monitoring systems", Document, IMdR, 2017.

[IMD 20] IMDR, "Big data in reliability", Document, IMdR, 2020.

[ISO 16] ISO 14224 Standard: 2016, "Petroleum, petrochemical and natural gas industries – Collection and exchange of reliability and maintenance data for equipment", ISO, 2016.

[LAN 94] LANNOY A., PROCACCIA H., *Méthodes avancées de traitement et d'analyse de banques de données de retour d'expérience*, Collection de la Direction des Études et Recherches d'Électricité de France, Eyrolles, Paris, 1994.

[LAN 01] LANNOY A., PROCACCIA H., *L'utilisation du jugement d'expert en sûreté de fonctionnement*, Editions TEC&DOC, Paris, 2001.

[LAN 11] LANNOY A., *Retour d'expérience technique*, Techniques de l'Ingénieur, SE1041, 2011.

[MAR 74] MARCOVICI C., LIGERON J.-C., *Utilisation des techniques de fiabilité en mécanique*, PSI, Technique & Documentation, Paris, 1974.

[ORE 15] OREDA, *Offshore and Onshore Reliability Data Handbook, 6th edition, Volume 1 – Topside equipment, Volume 2 – Subsea equipment*, OREDA, Norway, 2015.

[PET 99] PETTERSSON L., MOSS B., LANNOY A. *et al.*, "Handbook on quality of reliability data", An ESReDA working group report, Det Norske Veritas, 1999.

3

The Principles of Calculating Reliability in Level 1 PSAs

This chapter focuses both on the modeling (event trees and fault trees) used in Probabilistic Safety Assessments (PSAs) and on the mathematical justification of the calculations that are made from these models to arrive at an estimate of the frequency of unacceptable consequences (UCs) such as core meltdown of a nuclear reactor. This justification relies on a set of simplifying hypotheses of which the significance and consequence for the precision of the calculation are discussed. From this central calculation, we can deduce a ranking of the components according to importance and an estimate of the uncertainty due to an imperfect knowledge of the components' failure rates, which can be aids to decision-making for improvements in the design or maintenance of power stations. The end of the chapter proposes some future paths, in terms of modeling (with Boolean logic Driven Markov Processes [BDMPs]) and quantification (with the Initiator and All Barriers [I&AB] method). The subject of dynamic PSAs, mixing physical deterministic calculations and probabilistic calculations, is also mentioned.

3.1. Introduction

Despite the existence of normative or regulatory documents on PSAs [IAE 92, IAE 99, ASM 08, IAE 10, RFS 02], to this date there exists no text describing precisely the modeling techniques and general principles behind the calculations made. The most "operational" public document is probably given in [NUR 90a] since it contains figures (especially on rates of occurrence of initiating events), examples of models and calculation formulae but it gives no mathematical justification. This document is completed by [NUR 90b], which describes means of

---

Chapter written by Marc BOUISSOU.

taking account of fires and earthquakes. The absence of reference texts is compensated by knowledge that has been introduced over time in the most widely used calculation tools in the world of PSAs, that is, *RiskSpectrum*™, the reference tool in Europe, and its main American competitor *Cafta*. Documentation of these tools is centered on the user manual aspect and gives the specifications of the calculations made by the tool, without a high-level vision recalling the foundations of the calculation method. This chapter is limited to the domain of level 1 PSAs, whose main aim is to calculate the risk of core meltdown. By extension, it is also partly applicable to level 2 PSAs, which much resemble those of level 1 but distinguish a large number of consequences possible for event tree sequences. Levels 1 and 2 PSAs are still dominant in the nuclear industry to this day: for many power stations, level 3 PSAs have not been carried out. Furthermore, they call on much more varied techniques than those used for levels 1 and 2 and it would be difficult to give a unified vision of them. For example, document in [NRC 17] describes the American approach for levels 2 and 3 PSAs.

The calculation principles for level 1 PSAs rely on the WASH 1400 report by Professor Rasmussen from 1975 and have never been reworked. Research on tools and models from this pioneering work have improved calculation speed, tools' precision in the presence of large probabilities and the ergonomics of interfaces so as to be able to manipulate and process ever larger models, but the main principles have remained unchanged!

In view of the great complexity of power stations and their running procedures, approximations have to be made in the modeling to limit the calculation efforts needed. We will explain these approximations, "forced" by a principle of reality, from a mathematical point of view. PSA experts develop significant expertise, which is gradually being transmitted to newcomers in the domain. Yet there are many books of academic origin [RAU 20], or ones written by experts from industry [PAG 86, VIL 88, BOU 08a] describing fault trees and event trees (both types of model form the core of PSAs); the problem is that they describe these models without giving all the refinements used in the nuclear industry and illustrate them using very simple examples, which are insufficient for understanding PSAs.

In this chapter, we will first of all explain the anatomy of a PSA model: the way in which fault trees and event trees are articulated and used so as to provide a set of minimal cut sets (MCS), each one formed of an initiating event and of failures (failures on demand or in operation) of components of safeguarding systems activated to manage the initiator and return the power plant to a normal operational state. Fault trees are simple and well-known models, but event trees harbor a number of pitfalls that we will detail. We will give the theoretical foundations of the calculations made in PSAs, then we will explain how they are practiced in real and very complex models, while trying to outline the validity domain of the approximations used.

The end of the chapter is dedicated to two innovations in the domain of PSAs. The first concerns modeling: we will show how to gain both in clarity and in precision of modeling by replacing a composite model made of fault trees and event trees with a unique BDMP type model.

The second innovation is a quantification method that makes it possible to remove a number of limitations of the historic method. This method, called I&AB [BOU 16], greatly improves the relevance of the calculations made, while making it possible to re-use the existing models to a great extent. It has been available in *RiskSpectrum* software since the end of 2020. The I&AB method applies to classic models of PSAs and also to BDMPs.

3.2. The basis of all calculations: an exponential approximation

3.2.1. *The principle of exponential approximations*

We will first of all introduce some notations from Figure 3.1.

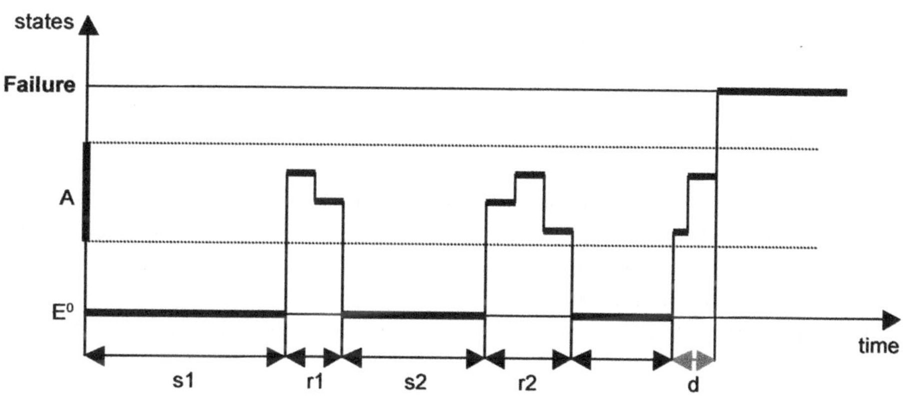

Figure 3.1. *Trajectory taking the system from the initial state E0 to a failure state. For a color version of this figure, see www.iste.co.uk/lannoy/nuclear.zip*

Figure 3.1 represents a system's "random walk", which takes it from an initial state to a failure state; for a reliability calculation, we can lump all the failure states into a single one, considered absorbing, because we are only interested in the date T of the first entry in this category of states. In Figure 3.1, A represents all the states that are different both to the initial state and the failure state.

If the system is totally repairable, such a trajectory consists usually of several journeys back and forth between E^0 and A, followed by a "single journey" to the failure state: so long as a state of catastrophe (in the context of a PSA, this will be core meltdown or radioactive release) has not been reached, it is possible to return the system to its normal operating state by repairing components that have broken down.

Exponential approximations consist of approximating the reliability of a totally repairable system using a function of the type $\exp(-\Gamma t)$, where Γ is a constant. Depending on context, we may be led to choose Γ so as to obtain a value as close as possible to the true reliability over a particular time range, or indeed to be pessimistic. This type of approximation was initially studied by Soloviev [SOL 83] and Keilson [KEI 86] who demonstrated limit theorems to their subject, nearly 10 years after the Rasmussen report! Keilson's approximation, which consists of taking as a value of Γ the inverse of the system's mean time to failure (MTTF) is potentially more accurate than no return to initial state (NRI) approximation that we will describe below, but it is not necessarily pessimistic. NRI approximation is always pessimistic and moreover, its parameter is easier to calculate.

We will give sufficient conditions for this approximation to be accurate; when this is the case, we can consider that there is an "equivalent failure rate" for the system, which evidently displays great practical interest. Thus, the overall result of a level 1 PSA is called the "core meltdown frequency" because it is not a probability; its dimension is inverse to a time, and it is usually expressed in per year, for a reactor. Supposing the independence of the reactors (which is not entirely true, because of external aggressions and resources shared with reactors situated on the same geographical site), one just has to add together the frequencies associated with reactors in a given country (there were 56 connected to the network in France in 2021) to obtain the rate of occurrence of a core meltdown accident in this country.

3.2.2. *NRI exponential approximation*

NRI exponential approximation [BON 92] involves considering that the time passed in degraded operating states A is small compared to the time passed in E^0, along a trajectory such as the one described by Figure 3.1; this is true for systems that are both totally and rapidly repairable. It should be noted that this reasoning is relevant for a reactor in its entirety, for a given "reactor state" (when powered, when shutdown for recharging, etc.), but it should not be used to study the individual reliability of safeguarding systems; such systems are in standby most of the time, and for them the "normal" state is a state where they are *not* fulfilling their intended function.

We can then estimate the instant of entry into failure states using a simple lower bound Tm: the sum of ν random variables written as s1, s2, etc., in Figure 3.1 (in other words, the sum of the times spent in E^0). It is enough to suppose that any entry into E^0 is a point where the process is renewed, and that the time spent in E^0 follows an exponential distribution with parameter Λ to obtain the expression of the NRI approximation. Under these hypotheses, ν is a binomial random variable, since each time that E^0 is left for A, there is a probability α of going into the failure state (without returning to the initial state, hence the name that we have chosen for this approximation) and 1 – α to return to E^0. Tm is therefore a geometrical sum of independent and identically distributed variables of parameter Λ, which makes it possible to assert that Tm has an exponential distribution, with parameter Λ α.

Since Tm is smaller than the time of entry into a failure state, this makes it possible to write the NRI approximation in its theoretical form, that is, the reliability:

$$\mathrm{R(t)} \geq \exp\left(-\Lambda\alpha t\right) \qquad [3.1]$$

Hereafter is the operative form in which this approximation is always used (an upper bound on the unreliability, which remains small compared to 1):

$$1 - \mathrm{R(t)} \leq 1 - \exp\left(-\Lambda\alpha t\right) \approx \Lambda\alpha t \qquad [3.2]$$

The shorter the time spent in A compared to the time spent in E^0, the closer this inequality is to an equality.

The article by [BON 92] contains, apart from this result, calculations of optimistic values of R(t) that make it possible, combined with the pessimistic value given above, to frame the reliability. It also contains a numerical example, for a Markov graph taken from a real study, for which these upper and lower bounds have been calculated. Within the framework of a PSA, calculation of an optimistic value would be too complex, but this is not very important because for a safety study, we wish above all to place an upper bound on the risk estimate.

NRI approximation requires calculating the two parameters, Λ and α, of the formula given above. Calculating Λ is very simple: it is the sum of the transition rates corresponding to the "initiator incidents", which draw the reactor out of its nominal state E^0. The calculation of α, in contrast, can quickly become very demanding in terms of CPU time if we proceed naïvely. This is where the approximations made in the PSA modeling will be involved.

3.3. The models used

To estimate the probability α of going directly from the initial state in which the power plant functions well to a state where core meltdown is inevitable, we use two types of models: fault trees and event trees. Rather than evaluating α overall, we calculate the probability after each initiator of moving to a UC, and we obtain the frequency of a UC (typically: core meltdown) by adding the frequencies associated with each initiator.

$$\Lambda\alpha = \sum \Lambda_i \, \alpha_i \qquad [3.3]$$

The general structure of a PSA model is given by its event trees. Each of them develops macroscopically the scenarios that can occur after a given initiator. An event tree sequence is formed of an initiator and a set of successes and failures of safeguarding system missions and sometimes successes and failures of the operator. Each sequence of the model is linked to a type of consequence: an acceptable consequence, core meltdown or radioactive release. These three categories of consequences are the only ones distinguished for a level 1 PSA. To move on to a level 2 PSA, the breakdown of consequences into categories needs to be refined much more, since they form sort of equivalence classes of the initial conditions of the sequences studied with the help of calculation codes modeling the physical phenomena at work after core meltdown, as far as fallout in the environment from radioactive products. The need for special processing for the initiators and the fact that there are several categories of consequences play a determining role in choosing an event tree rather than a simple fault tree. Fault trees serve to quantify the probabilities of safeguarding missions failing. The probabilities of human errors (actions not carried out or sometimes inopportune actions) are evaluated using specific methods described in Chapter 6.

3.3.1. *Event trees*

We will recall the representation of event trees in graph form succinctly here, along with the meanings that can be linked to it. Presentations of event trees that can be found in books rarely go beyond some general ideas that we will demonstrate below; for a more interesting presentation, because it integrates vast expertise on the use of this type of model, the reader can refer to [NUS 13].

3.3.1.1. *Simple event trees/event trees with dependencies*

Figure 3.2 represents a simple example of an event tree (real PSA event trees typically have a hundred or so branches).

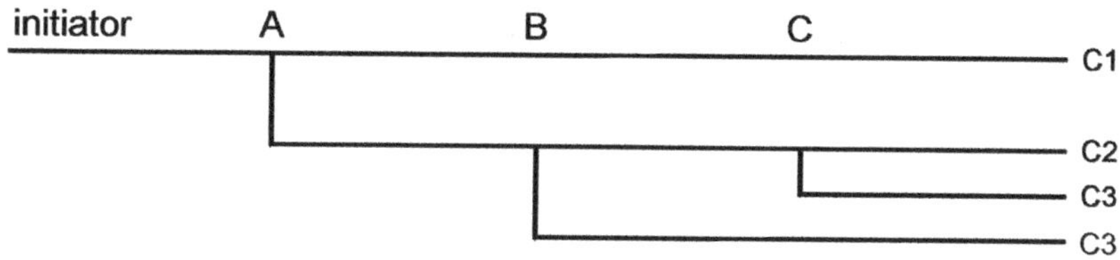

Figure 3.2. *A simple event tree*

This tree models the fact that following the initiating incident considered, we can implement three countermeasures (or barriers); these countermeasures correspond to the "function events", written at the top of the tree. If countermeasure A works (top branch), countermeasures B and C, which are not needed, are not triggered, and the consequences of the initiator are acceptable (C1). If countermeasure A fails, B and C must function to avoid UCs; we then arrive at a degraded state of the installation (consequence C2). If A fails and at least one of the two countermeasures B or C fails, UCs arise (C3).

First, suppose that countermeasures A, B, C are ensured by systems that do not share any components, and sufficiently varied that we can ignore failures with a shared cause that could affect two or three of them. So failure of countermeasures A, B and C would be independent events with probabilities written as pA, pB and pC, and the event tree supports an exact calculation of the probability of each type of consequence (conditional on the occurrence of the initiating event). Each end of one branch of the tree can be linked to the product of the probabilities of the branch points that lead to this end.

In addition, since two different branches are mutually exclusive, we can add the probabilities of the branches leading to a given consequence, as we do on our example for C3.

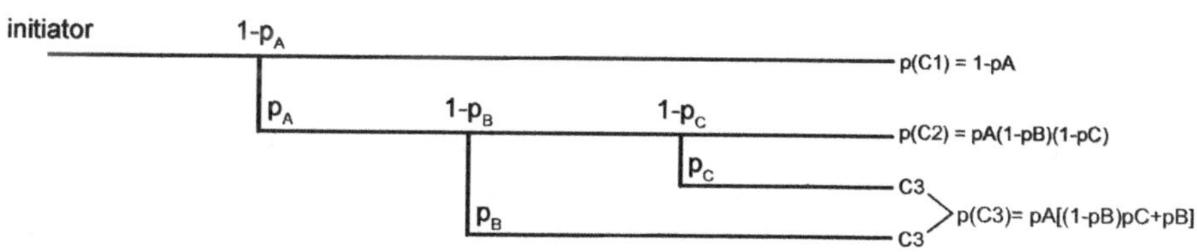

Figure 3.3. *Quantification of the event tree on Figure 3.2 (case with independent events)*

The example above involves binary branches, but there could very well be three branches or more for some function events; this generalization does not pose any problem so long as the branches correspond to mutually exclusive situations.

In reality, most often, function events correspond to the working of large complicated systems, whose failure is modeled by fault trees; failures of these systems are not independent because they share common support systems such as power supplies, or indeed because they are subject to failures with common causes.

To avoid any ambiguity below, it is useful to begin an aside here to give a definition of common cause failures (CCFs): CCFs are failures that can affect several components, simultaneously or during a system's mission, and which have the same cause, for example an error in design, manufacture, in the installation or assembly, a maintenance error, or an effect of the environment (fire, flood, earthquake, lightning). Two things distinguish a CCF from a functional dependency such as a shared power supply. A CCF causes a breakdown of the components in non-deterministic fashion, and when the initial cause has disappeared, it will be necessary to repair each component individually, which is not the case for a functional dependency. To simplify the models, we always consider that component failures due to a CCF are simultaneous; this simplification has the advantage of being pessimistic.

Dependencies between systems are the *raison d'être* of the method known as fault-tree linking, in which a Boolean function is associated with each branch of the event tree (called a sequence), this function being obtained from the conjunction of successes and failures encountered along the sequence. In our example, the branch leading to C2 would thus be associated with the Boolean function failureA and NOT failureB and NOT failureC. Each failure is detailed by a fault tree describing the combinations of basic failures that lead to this failure. But before explaining the link between event trees and fault trees (in section 3.3.2), it is useful to pause briefly at the "semantics" of event trees as they are used in PSAs.

3.3.1.2. *Ambiguity of event trees*

Event trees are models that some reliability specialists consider to be dynamic, as the vocabulary associated with them suggests, especially since a branch of the tree is called a "sequence". In PSAs, function events are placed as much as possible in an order corresponding to the triggering of safeguarding actions in time. Nevertheless, all the calculation methods classically used to quantify event trees are (for reasons of effectiveness) static, like for fault trees. This ambiguity arises in our opinion from two things: the fact that the graph representation of an event tree *obliges* us to choose an order for function events, which suggests the notion of a sequence even when there is no reason for there to be one, and the fact that when a branch is not opened at the level of a function event, this can be interpreted in two ways, one static and the other dynamic. So on the tree as shown in Figure 3.2, the success branch for A is not sub-divided at the level of B or C. This fact has two possible interpretations: either countermeasure A is tried at first, and, if it succeeds, there is no need to launch countermeasures B and C (we therefore have a real sequence, and a dependency from A to B and C – this is the dynamic interpretation), or A, B, and C are launched in any order and if A is successful, the fact that B and/or C fail is of no importance. As this changes nothing in terms of the probability calculation (as long as the static quantification method described in section 3.4.1 is used, since this does not involve time in the development of sequences), it is possible to choose the interpretation closest to reality. But it is clear that the event tree representation strongly suggests the dynamic interpretation. We have deliberately chosen our example so as to show that this representation can consequently be a handicap in understanding a model. In this example, B and C play perfectly symmetrical roles according to our hypotheses. Indeed, the event tree suggests that countermeasure B is tried before C, and that if B fails, C is not attempted. This, therefore, leads to a modeling dilemma: is it better to choose the event tree that reasonably suggests, but without the possibility of proof, that A is tried before B and C, and which creates a dissymmetry between the latter two by wrongly suggesting that B is tried before C, or the Boolean representation of consequence C3 which serves to support its calculation, that is, failureA and (failureB or failureC), which respects the symmetry between B and C but conceals the fact that A is tried before B and C? It is important to note that event trees are used primarily as a support for probability calculations, and should be accompanied by explanations as to how they have been created. This is the role of qualitative functional analyses, necessarily carried out upstream of their construction. These analyses, based on operation feedback and physical simulations, use a representation a little more convenient than event trees, "event sequence diagrams", (see section 3.7.1) that nevertheless convey the same type of ambiguity. It would be more satisfactory to have an unambiguous, "self-explanatory model". This would be the case opting for a BDMP, a model similar in graph form to fault trees, but which makes it possible to describe all sorts of dependencies between sub-systems and components. We will return to this in the last part of this chapter.

3.3.2. *Fault trees*

To take our little example above to its end, we consider the following very simple fault trees to explain the failure of countermeasures.

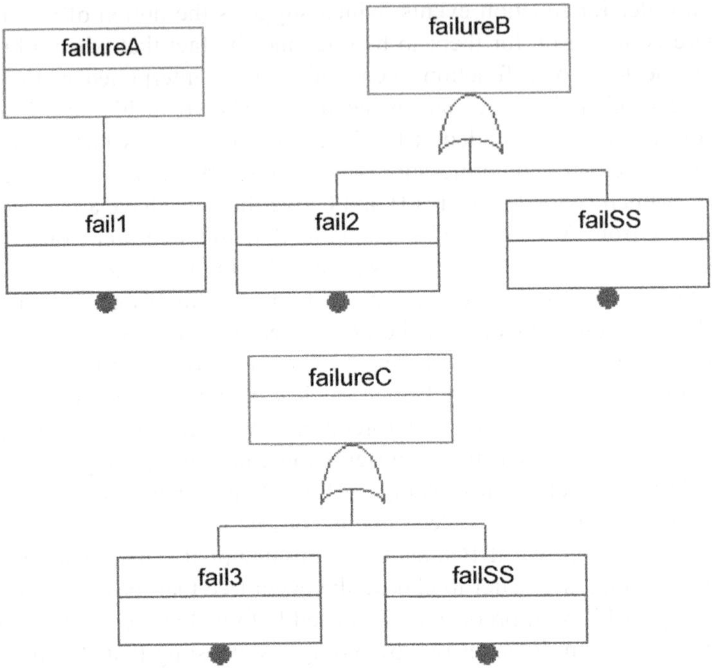

Figure 3.4. *Fault trees detailing failures of countermeasures of the event tree in Figure 3.2*

failSS is the failure of a support system (power supply or cooling for example) common to B and C.

The Boolean functions linked to consequences C1, C2, C3 when the failures of A, B and C are replaced by their details in the form of fault trees then become:

C1 = initiator AND NOT fail1

C2 = initiator AND fail1 AND NOT (fail2 OR failSS) AND NOT (fail3 OR failSS)

C3 = initiator AND fail1 AND ((NOT (fail2 OR failSS) AND (fail3 OR failSS))

OR ((fail2 OR failSS) AND NOT (fail3 OR failSS)))

Using the laws of Boolean calculation, the last two expressions can be re-written more simply:

C2 = initiator AND fail1 AND NOT fail2 AND NOT fail3 AND NOT failSS

C3 = initiator AND fail1 AND ((NOT fail2 AND NOT failSS AND fail3) OR

(NOT fail3 AND NOT failSS AND fail2))

We can see that even on a very simple example, we arrive at Boolean functions known as "non-coherent" or "non-monotonic", that is, in which the passage from FALSE to TRUE for a basic event can change the function from TRUE to FALSE (or vice versa). For example, consequence C1 equals TRUE if the initiator equals TRUE and fail1 equals FALSE. C1 changes to FALSE when fail1 becomes TRUE. In more concrete terms, this means that an additional failure can in some situations restore the system, or indeed on the contrary a repair can cause it to fail. A classic fault tree, built only with AND, OR and k/N gates, is necessarily coherent. But here, we have some negations necessary to represent the successful branches of the event tree. There are different techniques available in PSA calculation tools to manage these negations [EPS 05]: here, we will only speak of the one that is simplest to define mathematically. This is also the one that should be used so far as possible, but it can lead to substantial calculation efforts, which force us to use more approximative methods.

Coherent Boolean functions can be rewritten in a canonical form: an OR function whose inputs are the function's MCS. Each minimal cut set is itself an AND on a set of elementary faults. A cut set is a set of faults that are enough to trigger the undesirable event modeled by the Boolean function. It is called minimal if, as soon as one of its elements is withdrawn, the undesirable event is no longer true.

MCS are a very precious tool for analysts to verify the validity of their models. They are generalized by prime implicants for non-coherent functions, a prime implicant being the conjunction of a set of failures and of a set of failure negations. The problem is that the prime implicants are not very intuitive and are very numerous compared to MCS (if we compare a non-coherent function and its coherent envelope, which we define below). Another advantage of MCS is that a quantification can be made of the probability of the undesirable event from the probability of the MCS, while this is not true for prime implicants.

The points mentioned above mean that reliability specialists try as much as possible to turn to coherent functions, so as to be able to study them using their MCS.

This is why we are led to define the notion of a coherent envelope of a non-coherent Boolean function: this is the function that is obtained by keeping only the failures in the prime implicants. This envelope is a good approximation of the non-coherent function provided that all the probabilities of the basic events are "small" (say < 0.01 to give a concrete idea). In our example, this makes it possible to rewrite the different consequences in a much more comprehensible form:

C'1 = initiator

C'2 = initiator AND fail1

C'3 = initiator AND fail1 AND (failSS OR fail2 OR fail3)

All these approximations lead to overestimations of the probability. This is not a problem for those consequences that fall into the undesirable category since it is pessimistic.

3.4. Quantification of PSAs

Quantification has the aims of evaluating the frequency of "unacceptable consequences", calculating the importance factors of components or families of components, evaluating the uncertainty on the frequency of UCs from the uncertainty on the components' failure rates. These three aspects are discussed in the following.

Given the formula of the exponential NRI approximation given above, calculating the frequency of UCs amounts to calculating the probability of UCs, this probability being conditional on an initiator, from the MCS linked to this initiator.

The simplest method consists of placing an upper bound on the probability that a coherent Boolean function takes the value TRUE by the sum of the probabilities of its MCS: $\sum_{k=1}^{nb_{cuts}} \Pr(C_k)$. This method, known by the name of *rare event approximation*, has the advantage of not requiring the hypothesis of the independence of the basic events of the Boolean function. Indeed, it is not unusual in the models to find mutually exclusive events, such as indicator variables for different states of the reactor (when at full power, when in hot shutdown, shutdown for refueling, etc.), or indicator variables for different categories of flood, earthquake, etc. Adding together the probabilities of the MCS has the drawback of providing an estimate that can be excessively pessimistic (indeed it may even exceed 1, which is a bit surprising for a probability!). However, it is enough for *all* the probabilities of basic events to be low for this approximation to be of very good quality [COL 96]. A numerical application of the formulae of this article shows that for a number of basic events below 1000, it is enough for the probability of all the events to be less than 10^{-4} for the minimum cut set upper bound (MCUB) formula given below to improve precision by only 11% compared to the sum of the probabilities of the MCS. For the same relative improvement, if there are 10,000 basic events, it is enough for their probability to be lower than 10^{-5} (see section 3.6.1 for the number of basic events in a PSA).

On condition that the basic events are independent, we can use the MCUB formula: $1 - \prod_{k=1}^{nb_{cuts}}(1 - \Pr(C_k))$. This is the formula most used in practice, because it is more precise than the sum of the probabilities of the cuts and the additional effort to calculate it is negligible, but demonstrating the fact that it provides a pessimistic value requires the independence of the basic events. In fact, it is known that this formula gives an optimistic result when the model contains mutually exclusive events[1]. This is why some software specializing in PSAs make it possible to declare explicitly the mutually exclusive basic events and combine the two approximations cited above for the calculation.

There are various other methods of quantifying PSAs, which are more or less exact (especially on the question of considering the successful branches in event trees), which we will not detail here; some of them are described and discussed in [EPS 05]. The point to remember is that no "monolithic" method possesses simultaneously the advantages of providing a precise *and* pessimistic result and of scaling up to large models (especially where mutually exclusive events or successful branches with high probability exist). This is why software that specializes in PSAs uses combinations of these methods.

3.4.1. *Calculating the probability of UCs that are conditional on an initiator*

It is here that simplifying hypotheses become involved, which will make calculation possible, despite the problems of combinatorial explosion detailed in the following sections. Because of these hypotheses, the probability of each minimal cut set is calculated using a very simple formula that we will give further on. In a level 1 PSA, the two following undesirable events are usually grouped under the term "unacceptable consequence" (UC): core meltdown or early radioactive release, that is, release shortly after an initiator (e.g. steam generator tube rupture).

---

1 Here is a simple example. Suppose there are two exclusive events, A and B. This could be two minimal cut sets containing exclusive events. We therefore have $\Pr(\text{A OR B}) = \Pr(\text{A}) + \Pr(\text{B})$. Indeed, the formula MCUB gives: $\Pr(\text{A}) + \Pr(\text{B}) - \Pr(\text{A}).\Pr(\text{B})$.

We make the following hypotheses, which are the subject of international consensus:

– H1: After an initiator, within the space of 24 h (this time delay is called a scanning time), either UCs will have occurred or will be inevitable, or on the other hand, the incident will be under control and the UCs cannot occur until the next initiator. According to the calculation formula used, this amounts exactly to considering that where the incident is under control, the installation will return to its pre-initiator state. The scanning time can sometimes exceed 24 h for some initiators, for example if an inevitable event such as the emptying of a tank is foreseeable at 24 h. Generally, it is determined using physical simulations involving neutronics and thermo-hydraulics, which justify event trees. It never appears explicitly in probability calculations, where components' mission times[2] are involved: these mission times, necessarily shorter than the scanning times, correspond to the operation times needed to fulfill safety functions. In practice, in the vast majority of cases, the scanning and mission times taken equate to 24 h.

– H2: No repairs are made while managing an initiator (so for 24 h). Following H1, everything happens as if all the repairs, *including that of the initiator*, were achieved at 24 h.

– H3: After an initiator, all the safe-guarding systems are supposed to function in active redundancy.

– H4: There is no accumulation of independent initiators, that is, during management of an initiator, we take into account only the failures of the systems that form countermeasures to this initiator. Why this independence condition for initiators? Because for example we take into account the possibility of a loss of external sources (usually considered as an initiator) when this is a possible consequence of an automatic shutdown of the reactor.

– H5 (optional): For some initiators such as short-term network losses (of a few hours), we consider that if the initiator is repaired, the incident is under control (in other words, there is a return to the initial state). This makes it possible to take advantage of the fact that the initiator is repaired before 24 h.

H4 makes it possible to simplify models considerably, since while developing each event tree following an initiator, there is no need to examine what would happen if any other initiator incident (in general, more than a hundred are considered in a PSA) occurred before 24 h.

2 This should be understood as the functioning time for safe-guarding systems after an initiator, not to be confused with the mission time of the power plant in its entirety.

The other hypotheses make it possible to quantify MCS very simply. Thanks to H2, there is no need to know the components' repair time distributions. What will make it possible to limit the probability of the MCS, is H1 that limits the time during which a component of a minimal cut set can fail to 24 h. There are two types of failure: failure on demand, with a probability written γ, or failure in operation, with a failure rate supposed to be constant and equal to λ. H3 allows us to say that the mission time of each element of the cut set is equal to MT = 24 h, and not a random time dependent on other failures occurring such as in the case of standby redundancies (normal/standby operation), which leads to much more complex calculations [EID 16].

Finally, the probability (conditional on a given initiator) of a minimal cut set C_k carrying *m* failures on demand and *n* failures in operation is given by formula (3.4) that supposes the independence of the failures appearing in the cut set:

$$\Pr(C_k) = \prod_{i=1}^{m} \gamma_i \cdot \prod_{j=1}^{n} \left(1 - e^{-\lambda_j MT}\right) \qquad [3.4]$$

This formula gives both the unavailability of the cut set (the probability that it is present at instant *MT*) and its unreliability (the probability that it has happened before *MT*). This is due to the absence of repairs.

Let us recall that by adding the probabilities $\Pr(C_k)$ of the MCS, or by using the MCUB formula or possibly a calculation using a Binary Decision Diagram (BDD; see section 3.6.2), we obtain the term $\boldsymbol{\alpha}_i$ of the formula (3). According to this formula, it remains to multiply this conditional probability by the failure rate Λ_i of the initiator to obtain the frequency of the UCs consecutive to a given initiator.

Note that simplifying hypotheses can be pessimistic (H2, H3) or optimistic (H4). H5 is neutral, and H1 may be either optimistic, or pessimistic depending on whether the real repair times of the components are shorter or longer than 24 h. In total, it is not easy to say whether this calculation is pessimistic or optimistic.

3.4.2. *Calculating importance factors*

MCS are presented by calculation tools after sorting by decreasing probability (or frequency). Thus, it is possible to see quickly which the preponderant scenarios are, for which it may be necessary to take prevention or protection measures. In addition to this first approach, it is beneficial to calculate the importance factors linked to components or groups of components. Indeed, a heuristic way of optimizing the costs of modifying systems or operational and maintenance procedures is to act on the most "important" components. The ideal for reliability would be to achieve a "balanced design" in which no component has an importance markedly greater than that of the others. But this is very difficult in practice.

Boolean models show the advantage that they make it easy to calculate importance factors. It is important to note that many of them have both a probabilistic interpretation and an interpretation in terms of partial derivative. The importance factors of basic events are all calculated using the probabilities of UCs conditional on the failure/non-failure of a given component.

Since importance factors are numerous (there are many variants) and relatively redundant (we can regroup them into sub-sets giving similar results), we will cite only three of them: Birnbaum for the historical aspect [BIR 69], Fussell–Vesely [LAM 75] because it is intuitive and practical, and the risk increase factor [NUR 85] for its atypical side. The last two are the most used in the context of PSAs [BOR 01].

In the rest of this section, we consider the set of all the MCS linked to a given initiator as a Boolean function, and we write S the failure of at least one of the cut sets, and c that of component C. The formulae below are valid in a more general context that the one in which we are interested since they are applicable to unavailability (probability of failure at instant t) for any basic events, repairable or not. Since time is present everywhere, it is not explicitly written in the formulae.

Birnbaum introduced the first importance factor in 1969, which carries his name, and is also called marginal importance factor. It is defined as the partial derivative of the system's probability of failure with respect to the probability of failure of component C:

$$MIF(C) = \frac{\partial[\Pr(S)]}{\partial[\Pr(c)]} \qquad [3.5]$$

According to the formula of the total probabilities $\Pr(S) = \Pr(S|c)\Pr(c) + \Pr(S|\bar{c})(1 - \Pr(c))$, we can deduce:

$$MIF(C) = \Pr(S|c) - \Pr(S|\bar{c}) \qquad [3.6]$$

MIF also has a probabilistic interpretation: it is the probability that component C is "critical", that is, the probability that the system is in a working state such that if C breaks down from this state, the system breaks down. We note that MIF does not depend on $\Pr(c)$. In [ALI 17], the authors discuss (almost half a century after it was defined!) different extensions of MIF to the case of non-coherent Boolean functions (which requires redefining the notion of critical component) and present a new one.

Fussell–Vesely's importance factor (FVF) is calculated from MCS. This is the ratio of the probabilities of the union of the MCS containing C and the entirety of all the cut sets.

$$FVF(C) = \frac{\Pr(\cup\, Cuts\ containing\ C)}{\Pr(S)} \qquad [3.7]$$

This factor, comprised between 0 and 1, is all the closer to 1 the more the cut sets containing C are among the preponderant cut sets.

The importance factor called risk achievement worth (RAW) or risk increase factor (RIF) measures the increase in the unavailability of the system, which is due to the failure of a component. There is no interpretation in the form of a probability, since it can take unbounded values higher than 1.

$$RAW(C) = \frac{\Pr(S \mid c)}{\Pr(S)} \qquad [3.8]$$

A component whose RAW is high should be very reliable and should be subject to particular monitoring during use (maintenance), since its failure greatly raises the probability of the system failing.

The different importance factors do not always move in the same direction, this is why it may be difficult to identify the components to improve (by choosing more reliable ones, or by monitoring them more closely during maintenance). In practice, the RAW is used more for the reliability centered maintenance approach (optimization of preventive maintenance during operation), while the FVF serves rather to improve the safety of plants by identifying possible modification requirements. It should be noted that, since the formulae of importance factors are defined for one instant *t*, classification of the components can vary during the course of the system's mission. However, in the context of PSAs, we are generally interested in a single time value: *MT*, the mission time used to quantify the consequences following an initiator. To these theoretical difficulties are added the risks of poor precision in calculating importance factors when we use a selection of MCS (the most probable) and not the whole list. Indeed, this is nearly always the case.

3.4.3. *The uncertainty calculation*

Operation feedback may be non-existent or insufficient to give precise estimators of failure rates (see Chapters 2 and 5). This is why it is important to measure the effect of this uncertainty on the overall results of the PSA.

The principle consists of considering the failure rates as random variables in place of fixed values. The probability distribution chosen for each rate represents, through its greater or lesser spread, the uncertainty on these data, an uncertainty that depends on the number of observed failures considered in its calculation. To obtain the uncertainty on the overall result of the PSA, calculation tools make a Monte Carlo simulation: they calculate the overall results obtained from a large number (typically several thousands) of sets of failure rate values taken at random according to their distributions. They can thus produce a histogram of the results obtained, and also a confidence interval for the result.

The result of the PSA can be seen in a first approximation as the sum of a large number of MCS probabilities. There are therefore compensation phenomena that make the uncertainty on the overall result smaller than that on many components.

It would take far too much calculation time to carry out the complete calculation of the MCS for each draw. So, these repeated evaluations are made only on the MCS that have been retained, because their probability is above the truncation threshold, in the reference calculation (i.e. using the best estimated values of each of the failure rates). This clearly degrades the precision of the uncertainty calculation, but it seems impossible to do otherwise.

In [BOR 03], the subjects of importance factors and uncertainty are brought together. It explains why it is important to consider in Monte Carlo runs the dependency between basic events due to the fact that they share common parameters (e.g. all pumps that are identical and maintained in the same manner have the same failure rate). It also shows that the components at the top of the list in the classification by importance factors are not necessarily those that contribute the most to uncertainty in the final result. This comparison is not trivial, to the extent that for importances, we classify basic events, while for uncertainty, we classify the parameters, which are fewer. This article relies on the numerical example of what could be called a "micro-PSA", with a single initiator (a loss of coolant on the primary circuit) and 44 basic events involved in barrier failures.

3.5. The question of the level of detail

The main obstacle to quantifying UCs from a set of MCS is the breaking down of the probability of failure sought on a very large number of cut sets of contributions that are pretty much equal. It is easy to build a pedagogical example presenting this particularity, but very fortunately, real systems only rarely possess this. In fact, they very often have weak points; the probability of failure due to these weak points is then so preponderant that it allows us effectively to place an upper bound on the probability of unexplored cut sets and to obtain a precise result after all.

Any dependability model of a complex system, to the extent where it should theoretically be capable of considering all the possible combinations of failures of the system's components, is subject to a combinatorial explosion phenomenon. Depending on the types of models used and also the calculation methods chosen for a given type of model, there are different ways of compromising with this major constraint.

But finally, we are always led to choose a trade-off between too fine a level of detail, which causes an explosion of the calculation methods, and too coarse a level of detail, which provides imprecise results or ones so aggregate that it is not possible to distinguish those of interest.

Let us take a basic example: suppose that the loss of a system S can be explained by the conjunction of the failures of two independent sub-systems X and Y, and that each of these sub-systems is formed of n components.

Depending on whether we take the sub-systems or the components as the level of detail, we will have 1 or n^2 combinations of failures explaining the loss of S. If we are interested only in the unavailability of S, the simplest model suffices. It makes it easier not only to build and evaluate the model, but also to use the operation feedback on which the model relies. Indeed, it is possible to estimate the unavailability of each of the subsystems globally (i.e. without having to attribute it to the different components) and avoiding the problem of estimating a failure rate in the absence of observed failure (see Chapter 5).

The only drawback of this simple model (with a single minimal cut set) is that it does not allow us to distinguish the contributions of elementary components of sub-systems X and Y. It is therefore impossible to deduce from it, for example, which components need to be improved to reduce the unavailability of S as effectively as possible.

If n is large, we see immediately that it is useless to show all the n^2 failure scenarios to the analyst. Only those that make a significant contribution can be of interest. A selection of those scenarios that are preponderant in terms of probability is therefore indispensable in any even slightly complex model. This selection *must* be accompanied by an upper bound of the error made, neglecting the other scenarios. In fact, without doing this, it would be enough to refine a model to reduce the probability of the undesirable event! This phenomenon is easy to explain on the example of the X – Y system. If each of the sub-systems is formed of 100 components, all independent, whose probability of failure is 10^{-4}, each sub-system's probability of failure is around 0.01, and so the system's probability of failure is close to $0.01^2 = 10^{-4}$. By taking the finest level of detail, we have 100^2 MCS, each with a probability of 10^{-8}. By choosing 10^{-6} as a threshold for eliminating cut sets, the system's probability of failure is estimated at 0! This pitfall is avoided, if we consider that at worst, we would have had a probability equal to the number of MCS (which can be estimated without making them explicit) multiplied by the probability threshold, which, with the numerical values above gives an upper bound for the system equal to 0.01. If this figure is deemed unacceptable, it shows that the threshold of 10^{-6} is not satisfactory and that it is necessary to take the calculations further. In reality, with models of varied logical structure and large differences between the failure probabilities of the least and most reliable components, calculating an upper bound is much more complex but still possible.

The level of detail usually used in PSAs is such that the basic events are the components' main failure modes. Here, the "component" can range from a simple manual valve to a sometimes quite complex sub-set such as a pump with its cooling circuit, or a diesel generator. With the aim of optimizing the models' calculation times, one might imagine having a variable level of detail, by creating macro-components regrouping components independent of the others. But this would go against data traceability, against the use of databases with standardized content, and this could lead to concealing component losses, which could become significant in PSAs dealing with aggressions (see section 3.6.2). Anyway, the first stage of calculations made by the tools is the identification of the modules in the fault trees, which means that most of the relevant groupings are made automatically. In addition to optimizing calculation times, these groupings make it possible to avoid the pitfalls described in the previous section.

3.6. Practical problems: model size, high probabilities

With increased system complexity, the enrichment of operation feedback that pushes us to consider ever more types of failure and the extension of PSAs' domain of coverage, the PSA analyst is confronted with two types of difficulty: the combinatorial explosion of processing and the need to consider high probabilities (0.1 and more) during calculation of the conditional probability of UCs directly following aggressions such as fire, internal flooding, earthquake or external floods.

3.6.1. *Model size and combinatorial explosion*

At EDF, the first PSAs involved 900 and 1300 MW plants. With the level of detail mentioned at the end of the previous section, their "volumetrics" can be described by the following figures (model example for a 900 MW plant):

– 7,683 basic events;

– 450 CCF groups;

– 818 house events (these are the fault tree leaves to which it has been decided to give the value TRUE or FALSE following the tree structure that it is desirable to retain to model the failure of this or that system mission);

– 12,612 logic gates in the fault trees. It is scarcely possible to give a number of trees since a tree in fact contains a number of others, corresponding to the different sets of values that house events can take;

– 272 pages of event trees; the number of branches for the vast majority of pages ranges from around 10 to around 30. There are more pages than event trees, since for reasons of readability, we use redirections between pages: what seems to be the end of a branch is prolonged. In total, this makes 4,034 sequences for the complete PSA (1 event tree branch = 1 sequence).

The complete calculation for the core meltdown consequence for such a model takes around 1 minute on a standard PC, with a probability threshold giving a satisfactory precision. Once the preponderant MCS are obtained, they can be used to make an uncertainty calculation and/or to calculate importance factors. All these operations are also carried out in a few minutes.

Things have become much more difficult with the new EPR reactors. In fact, these reactors are more redundant than previous generations. For example, power supplies are distributed in four divisions, each of them including a diesel generator. Two additional diesels can replace the main ones in case of failure. This high degree of redundancy explodes the number of MCS to be considered, and hence the computation time. The solution consists of cutting the PSA model down into several parts, which will generate disjoint sets of MCS; this is the case during an initiator by initiator processing. It is then enough to add (thanks to the fact that the sets of MCS are exclusive) the partial results obtained to have an overall result. The Andromeda tool from EDF (see section 3.9) provides diverse aids for the partitioning of calculations, which are especially useful for making good use of several machines performing calculations in parallel [HIB 20].

3.6.2. *Fire, flood and earthquake PSAs: the problem of high probabilities*

Fire and flood initiators (whether the flood originates inside or outside the power plant) or earthquake initiators raise a particular problem: the probabilities of components failing, conditional on these events can be high: 0.1 or more.

As the levels of uncertainty linked to these initiator event categories are very different, it is usual practice to begin with a PSA of internal events, then to extend this to fire, flood and earthquake in separate studies. Although this approach is not used in France, we point out article [BAN 17] which proposes a method for calculating an overall risk using aggregation of risks associated with different uncertainties. In the United States, the question of aggregation is addressed from the angle of risk-informed decision-making. In [FER 21], the authors give the principles for presenting the results of risk analyses and the possible choices to a decision maker in different situations such as optimization of maintenance or modifications to an existing power plant.

For floods and earthquakes, different levels of severity are considered: flood levels are associated with return periods (10 years, hundred years and thousand years) earthquakes are characterized by the intensity of their effects (in terms of maximum acceleration, frequency) depending on their magnitude on the Richter scale, the distance of the power plant from their epicenter, etc. The models produce MCS of which one of the basic events is an indicator variable of the level of severity considered.

For the highest levels of severity, some components' probability of failure can be as high as 1. More precisely, in the case of a flood, it is considered that components under water fail with a probability of 1. For an earthquake, "fragility curves" are used, which give components' and structures' probability of failure depending on the maximum acceleration on the ground; these curves are estimated using post-earthquake operation feedback describing the behavior of similar equipment during earthquakes and equipment design rules (taking account of the earthquake or not), physical tests or simulations.

Events with a probability of 1 can be replaced by basic events that are always true, which makes it possible to simplify the models from the perspective of logical structure. But for the others, we may be very far from the conditions making it possible to guarantee that the sum of the probabilities of the MCS is a good approximation. For a given set of MCS and on condition that the basic events that figure in them are all mutually independent, the best quantification method uses a BDD [EPS 05, ZVO 18]. The results associated with sets of MCS of different severity should quite simply be added, since these sets of MCS are mutually exclusive.

For fire PSAs, we begin with a fairly simple analysis called "volumes of fire" considering that all the materials situated within the volume of fire containing the source of fire are lost, then a more in-depth analysis is made using a fire spread calculation code for the preponderant terms. Some fires only cause the loss of safe-guarding systems and others can both cause an initiator event and the loss of some barriers.

1) The volumes of fire are located (they are bounded by fire doors). The impacts of the total loss of components in a volume of fire are considered. We take as fire frequency of a volume of fire the sum of the frequencies associated with the potential sources contained in this volume.

2) For volumes of fire giving the highest probabilities of a UC, we model the geometry of the locations and materials that they contain in a fire spread calculation code to see more precisely which components are affected depending on the starting point of the fire. This amounts to breaking consideration of the volume of fire down into several cases.

3.7. "Cousin" models of PSA models

3.7.1. *Event sequence diagrams*

There are representations that are a little easier to build than event trees, and which can eventually be transformed automatically into event trees: these are event sequence diagrams (ESD) [SWA 99]. It is enough to show the equivalent in this formalism of the event tree from Figure 3.2 to understand what it is. In an ESD, the labels of the barriers to be considered after the initiator are in the boxes, of which the right-hand outcome is the favorable result, while the outcome below is a failure of the barrier. By exploring all the paths starting from the initiator in this graph, we find all the sequences of the event tree. One important advantage of ESDs is that it makes it possible to factorize shared endings of sequences, whereas they need to be repeated in event trees. Here, this is done only for the last two sequences. ESDs are used in PSA qualitative sequence analyses (QSAs), operational studies that rely on physical simulations of transients triggered by the initiators; they are then translated into event trees for probabilistic quantification.

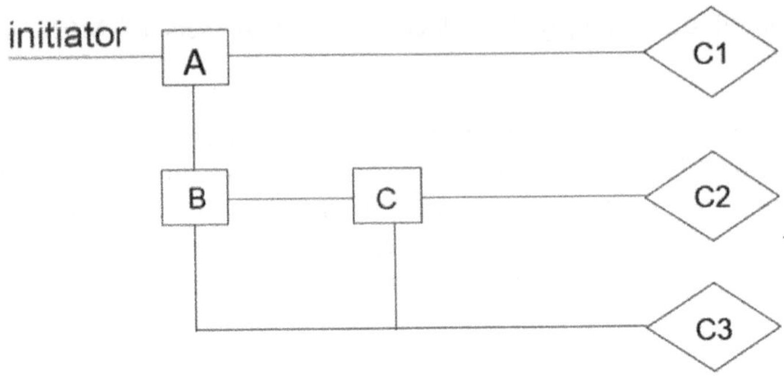

Figure 3.5. *ESD equivalent to the event tree in Figure 3.2*

3.7.2. *Bow tie diagram*

The "bow tie" model takes its name from its general appearance in graph form, since it is formed of a "lying" fault tree (drawn with the leaves on the left and the top event on the right) and of an event tree. This graph view is convenient so long as we stick to models with little detail.

A unique, feared event called a "pivot" or "central" event serves as a junction between the two sub-models: it is both the top event of the fault tree and the initiator of the event tree. The choice of this event is clearly determining for the whole model. It generally represents the boundary between the two types of measures that can be taken to reduce risks: prevention (i.e. decreasing the probability of occurrence) on the left and mitigation (or minimization of consequences) on the right. Typically for a refinery, this could be the start of a fire or an explosion. In the fault tree part, there would be represented all the barriers put in place to avoid the fire/explosion, whereas in the event tree part there would be represented barriers, such as geographical distancing and fire walls to counter its spread, the intervention of the fire brigade, etc.

The bow tie diagram is used mainly in hazard studies imposed by the European directive SEVESO [SEV 12], for chemical, gas or oil installations. In [QUI 20], the authors compare use of the bow tie diagram with that of the PSA method specific to the nuclear domain.

3.7.3. *Boolean logic-driven Markov processes*

Originally, BDMPs were designed to offer an easy specification for dynamic models of type continuous time Markov chains (CTMC), with a large number of states [BOU 03]. BDMPs have a graphical representation very similar to fault trees, so they are much easier to master than Markov graphs or stochastic Petri nets (and their many variants). They have been extended to the field of security. Thanks to the unification of the concepts used in safety and in security, BDMPs make it possible to construct models containing both these aspects, so as to be able to evaluate overall the risks linked to a system, and to be able to take good decisions in cases where an antagonism exists between the measures destined to ensure safety and security [KRI 14]. Software from the KB3 tool platform [BOU 05], developed by EDF, make it possible to build and process BDMPs easily. They are well adapted to modeling complex systems thanks to their mechanism known as "relevant event filtering", which considerably reduces the size (in terms of the number of states, transitions and sequences) of the Markov processes they specify. To give an idea of this filtering principle in a few words, we can simply say that it makes it possible to inhibit failure of the leaves when these are only involved in sub-trees of the BDMP where the top already has the value TRUE because of previous failures. This filtering on the model is equivalent in the physical world to not considering additional failures in a subsystem that has already broken down. In the vast majority of cases, this is more realistic than considering them; anyway, it is possible not to apply the filtering, either locally to a sub-tree or overall, when this hypothesis is not the right one.

It turns out that in addition to the advantages above, BDMPs can form an alternative representation to event trees, which removes the drawback described in section 3.3.1.2, as well as others, detailed in the article [BOU 08b]: sensitivity to the order of function events, excessive size making it necessary to truncate branches and so facilitating errors, not taking account of time dependencies. It is important to note that use of a BDMP does not oblige us to abandon the effectiveness of Boolean methods when we reach quantification. The BDMP in Figure 3.6 models *exactly* the hypotheses on the system's functioning from section 3.3.1.2, thus removing all ambiguity; the trigger (dotted red arrow) of this BDMP makes it possible to see at first glance that *both* barriers B and C are triggered *simultaneously* on the loss of A. In addition, the BDMP gives *clearly symmetrical* roles to B and C. If we wish to model the fact that barrier B is tried first, and that C is tried only in the case where B fails, it would be enough to add a second trigger going from B to C in the BDMP in Figure 3.6.

So that this model can serve as an aide to calculating the probability of consequences C2 and C3, it is enough to complete it using a state observation function. This can be defined in graph form by extending the BDMP. In fact, consequence C3 coincides with the realization of the AND gate at the top; we can define consequence C2 using the small Boolean (non-coherent) function C2= failureA AND NOT (failureBackup) shown dotted in Figure 3.6. Thus, although a BDMP necessarily has a monotonic structure function, it can serve to determine the non-monotonic Boolean functions associated with degraded states of the installation, such as C2 in our example.

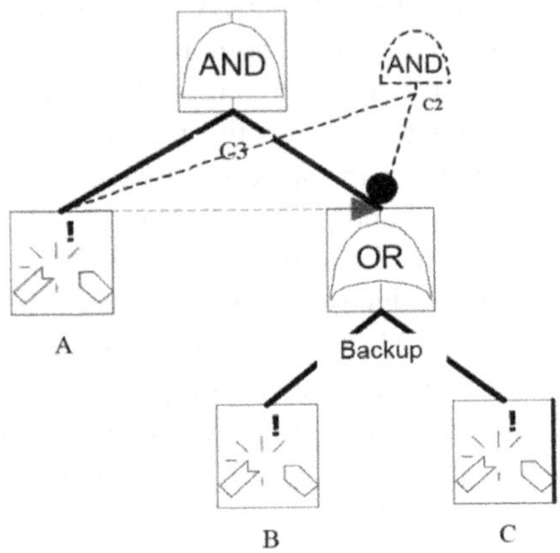

Figure 3.6. *Alternative to the event tree in Figure 3.2. For a color version of this figure, see www.iste.co.uk/lannoy/nuclear.zip*

By applying one of the methods described in [BOU 20a] to this BDMP, we obtain the "minimal contents of the sequences[3]", which leads to the two types of consequence described in the model:

– {failureA, failureB} and {failureA, failureC} for C3;

– {failureA} for C2.

Note 1: The "minimal contents of the sequences" is a monotonic function, we therefore lose track of the fact that to arrive at C2, it is necessary that neither failureB nor failureC occurs. This is not a serious problem from the perspective of placing an upper bound on the probability of C2.

Note 2: If A, B and C are not basic events but correspond to barriers whose failure is detailed by fault trees, it is enough to have a complete model to replace the leaves of the BDMP in Figure 3.6 with the fault trees in question, since they can be considered as sub-BDMPs without trigger. We thus use a unique formalism for the complete model (see Figure 3.8 for an example).

3 This is the generalization of the notion of minimal cut set to dynamic systems. See [BOU 20a] for a formal definition and varied examples.

The operations that we performed in Figure 3.6 can be generalized to a "BDMP + Boolean observation functions" set as an alternative to any event tree [BOU 20]. After transformation of the BDMP + observation functions set, we obtain a fault tree in which the different events at the top correspond to the different consequence categories of the event tree. By calculating the MCS linked to these events at the top, we obtain the minimal contents of the sequences leading to the different consequence categories for the installation modeled. We can then estimate the probabilities of these consequence categories using either a simple calculation corresponding to formula (4) for a calculation without repair (in view of hypotheses H1 to H3, it would be more accurate to say with a uniform repair time for all the failures), or the I&AB calculation formula (see section 3.8.1) for the case where we wish to consider the real repair times.

Here is a summary of the proposition made in this chapter to replace event trees with BDMPs:

– A BDMP (possibly supplemented with Boolean observation functions) makes it possible to model exactly and very readably the dependencies and independences between barriers by removing all ambiguities due to representation using event trees. It makes it possible to draw out clearly the symmetries that may exist between the roles played by different systems.

– This BDMP can serve:

- either directly, with a Markovian analysis tool, for a precise quantification of a consequence category of the event tree;

- or to determine a set of coherent Boolean functions able to serve as an aid for an approximate quantification, as it is usually made using PSA models.

A remark on modeling CCFs: to model CCFs "by hand" using the basic concepts of BDMPs can prove laborious if there are both CCFs on demand and in operation, affecting more than two components. But the BDMP knowledge base used with the KB3 tool has contained the notion of "CCF group" since 2020. It is enough to declare a CCF group with its numeric parameters and the list of leaves on which the CCF can act so that all the failure combinations, two by two, three by three, etc., are considered in the dynamic version of the model. And if it is transformed into a Boolean fault tree to use the I&AB method detailed below, it is enough to transfer the information on the CCF groups to the *RiskSpectrum* tool, exactly as in a classical PSA model.

To this day, even though BDMPs have been exploited in many studies of real systems, they have not yet been used on a large scale as an alternative to event trees. An experimentation needs to be carried out to get some feedback before proposing modifications to software tools (especially *RiskSpectrum* and *Andromeda*) that will make it possible to integrate them entirely into the process of building PSAs.

3.8. How can we improve the precision of classic PSAs?

3.8.1. *Principles of the I&AB method*

With the I&AB method [BOU 16], we can use the same set of MCS as in the classic PSA model[4], but the way of quantifying the cut sets has been changed radically. There is no longer any need for a mission time, since what makes it possible to limit the probability of conditional failure associated with an initiator is the repair of the initiator and other components appearing in MCS. The I&AB method gives, using an analytic formula, a good approximation of what would be provided by the quantification of a small Markov graph representing the behavior of elements of the minimal cut set after the initiator. Figure 3.7 represents such a graph, for a minimal cut set corresponding to the failure of three components, written 1, 2 and 3. State A is reached just after the initiator at time t_0. Repairing this initiator leads to an absorbing state representing a safe state. The conditional probability of realization of the cut set after the initiator is the probability of reaching the failure state written "fail", rather than going to a safe state, on an infinite time horizon.

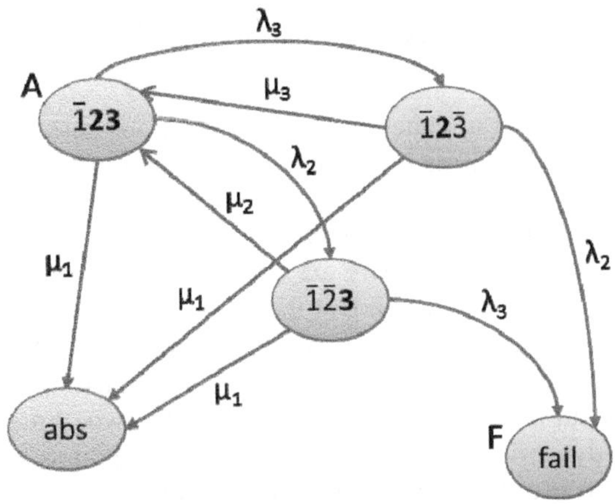

Figure 3.7. *Theoretical model approximated using I&AB calculation formulae. For a color version of this figure, see www.iste.co.uk/lannoy/nuclear.zip*

4 We use the term "classic PSA" to designate a PSA built and quantified according to the principles described from the start of this chapter to the end of section 3.7.1.

Calculating the conditional probability of realizing the cut set following its initiator relies on a very complex mathematical formula demonstrated in [BOU 16] and [BOU 18] but calculation of which is nonetheless much faster than creating the Markov graph matrix and quantifying it on an infinite time horizon. The simplifying hypotheses allowing this calculation to be made are the following:

– Failures on demand occur (or not) at t_0. If they are repairable, at the repair of any one of them, the cut set becomes definitively impossible to realize. This hypothesis fits well the case of unavailability due to preventive maintenance, which is revealed when a component is needed following an initiator. For other cases, it is more sensible not to consider repair to ensure a pessimistic calculation.

– Failures in operation can occur and be repaired indefinitely, so long as all the elements in the cut set do not fail simultaneously. In the example in Figure 3.7, this corresponds to a circulation between the three states shown in blue.

We therefore see that among the simplifying hypotheses of the classic method, we retain H3 – the functioning of the components (elements of a minimal cut set except the initiator) in active redundancy, we "replace" H1 and H2 by considering repairs, we retain H5 (repair of the initiator leads to a safe state). But I&AB makes it possible to have very long repair times; for example, repair of the grid when it is affected by climatic conditions that bring down pylons can take around 10 days.

As for H4 (the non-accumulation of initiators), it is difficult to do any better if we continue to build models of sequences by hand in the form of event trees. However, if the initial model is a BDMP, we can consider the accumulation of two initiators or more without any problem since the method described in [BOU 20] automates the process of obtaining the minimal contents of the sequences from a model that implicitly contains the description of all the possible sequences, including those with accumulations.

3.8.2. *What gains does I&AB allow?*

The first gain affects the precision of the ranking of components by decreasing importance. In fact, examples of calculation carried out on complete PSA models [BÄK 18] have shown that, provided the average repair times of the different components do not stray much from 24 hours, the overall figure for core meltdown frequency changes little when we move from the classic method to I&AB; this is easily explained by a compensation phenomenon between the MCS whose probability decreases and those for which it increases. On the other hand, the ranking of components can be changed significantly [BOU 20b]. Indeed, this ranking is important, since it is a driver for making decisions to modify the design or operation and maintenance of systems to improve safety.

The second gain provided by I&AB is the possibility of modeling satisfactorily the accidental scenarios linked to the loss of cooling of the spent fuel deactivation pool. Because of the large quantity of water it contains, the pool provides a thermal inertia making it possible to go through periods of total loss of cooling. This fact is incompatible with simplifying hypotheses H1 and H2. In fact, after a (highly improbable) accident scenario causing the total loss of cooling in the pool, even if it is re-established within 24 h, this is a degraded situation where the water boils in the pool and its level has started to drop: it seems very optimistic to say that if the fuel has not melted, everything happens as if there was a return to the initial state. With the extension of the I&AB method described in [BOU 18], we can consider two types of deterministic times: grace periods such as that due to the inertia of the pool and some failures such as those due to the exhaustion of a battery or water reserve. Because of these two possibilities, the precision of the evaluation of the probability of the fuel no longer being immersed in the water is much improved, and this could be shown by a comparison using a Monte Carlo simulation of a model much closer to the physics. This type of comparison has been made at EDF starting from a BDMP type model, but also by NKS (Nordic nuclear safety research). The report in [NKS 21] makes a comparison between I&AB in the context of a PSA of reduced but realistic size with the classical method and two others. These two other methods are as follows: the classical method in which a specific treatment is made of the preponderant cut sets by adding to them a basic event whose probability is supposed to take account of the possibilities of repair, and a method in which the preponderant cuts are recalculated using a Monte Carlo simulation of a dynamic model taking the physics into account in a simplified fashion.

The improvements proposed in the context of I&AB are evolutions that remain reasonable in terms of computation power needed and they could be applied in the short-term, using for the most part already existing models.

3.8.3. *Numerical application of I&AB*

The following example allows us to illustrate both the benefit of using a BDMP type model and the gain in precision provided by I&AB. Qualitatively, the BDMP from Figure 3.8 corresponds to the combination of the event tree described in section 3.3.1 and the fault trees from section 3.3.2. To be able to quantify it, it remains to choose reliability data of the same order of magnitude that we find in PSAs. For repair times, there are two sets of data, presented in Table 3.1 (the "true" data), and a uniform 24-h value applied to all the events, including the initiator (to get closer to the hypotheses of the classical PSA method).

Basic event	Failure rate/h	Average repair time, h
Initiator	1.E-6	200
fail1	1.E-3	4
fail2	1.E-3	50
fail3	1.E-3	20
failSS	1.E-3	10

Table 3.1. *Reliability data*

Table 3.2 gives the calculation results for two variants of the BDMP: the "standby redundancy" variant corresponds exactly to the BDMP from Figure 3.8, whereas the "active redundancy" variant corresponds to this same BDMP, with the lower trigger removed. This lower trigger indicates that the back-up barriers B and C are required (and so can fail) only if barrier A fails. Given that this barrier is repaired very quickly, in 4 h, it is easy to guess that this should do a great deal to lower the final result, that is, the frequency of the top event of the BDMP occurring, corresponding to consequence C3.

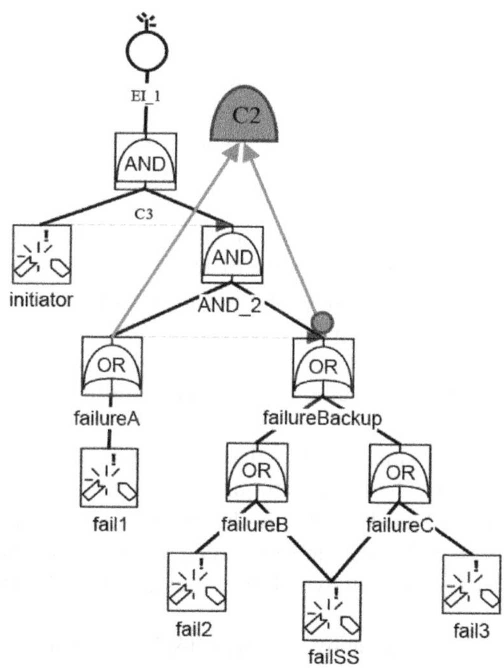

Figure 3.8. *BDMP replacing the models from Figures 3.2 and 3.4 (consequence C2 is defined by the observation function in blue). For a color version of this figure, see www.iste.co.uk/lannoy/nuclear.zip*

Table 3.2 contains the calculation results, numbered from 1 to 8 to facilitate the commentaries that follow. The most precise results are calculated using the Figseq tool [BOU 05]; they correspond to the NRI approximation calculated on the Markov graph specified by the BDMP. We see that the standby redundancy is better than the active redundancy by a ratio of around 7 with the "true" reliability data set (results 3 and 7), but this ratio drops to 2 with uniform repair times of 24 h (results 1 and 5).

Active redundancy				Standby redundancy			
Figseq, all repairs in 24h	PSA classic method (mission time 24h)	Figseq	I&AB	Figseq, all repairs in 24h	PSA classic method (mission time 24h)	Figseq	I&AB
1.61E-09	1.65E-09	1.43E-08	1.54E-08	8.24E-10	1.65E-09	2.31E-09	1.54E-08
1	2	3	4	5	6	7	8

Table 3.2. *Hourly frequency of consequence C3*

The classic PSA and the I&AB methods do not differentiate between active and standby redundancy: this is why result 2 is equal to 6 and 4 is equal to 8.

The classic PSA method gives almost the same thing as the precise calculation made with Figseq on the model in active redundancy and with uniform repair times of 24 h, which is not surprising if we refer to hypotheses H1, H2 and H3 from section 3.4.1.

The I&AB method always gives a pessimistic result: result 4 is higher than 3, and result 8 is higher than 7. This property of I&AB of being pessimistic is general so long as we consider failures on demand as non-repairable, but this can lead to an excessive conservatism on some MCS.

The classic PSA method can give an optimistic result (see results 6 and 7), or even a very optimistic result (see results 2 and 3). This is clearly due to the initiator's long repair time, since most of the other repair times are lower than 24 h. To ensure the pessimism of the PSA method, we would need to consider a scanning time much longer than 24 h. Taking 200 h, the initiator's average repair time (a time that has a probability of 0.37 of being exceeded by the real repair time), the PSA method gives a result of 8.18E-8/h, which becomes excessively pessimistic.

The conclusion of all the above is that I&AB effectively and significantly improves the precision of PSAs, while still making it possible to guarantee a pessimistic result compared to reality, which is not the case with the classic method (see Table 3.2). Calculations by sequence exploration in the Markov chain specified by a BDMP (the method used by Figseq) cannot scale up an entire PSA, unlike the

I&AB method [BÄK 18]. Impediments on the use of I&AB in an industrial context are as follows: its very recent availability in the *RiskSpectrum* tool (mid 2020), the need for additional data (repair times in incident contexts) and the work of revising existing models. Even though, *in fine,* the modifications are not numerous, we cannot escape verification of an entire PSA model when we wish to adapt it for I&AB.

3.9. A line of research: "dynamic PSAs"

Classic PSAs rely on two types of model: fault trees and event trees. Automation of the construction of fault trees from physical system diagrams has already been practiced at EDF using the KB3 tool for a number of years [PIL 99]. In contrast, the construction of event trees remains a very complex task and requires data and hypotheses, if possible realistic, issuing from different domains: neutronics, the physics of thermohydraulic monophasic and diphasic systems, the control logic, incident and accident conduct procedures. There are calculation codes such as CATHARE (a code developed in collaboration with the CEA, EDF, AREVA and the IRSN) capable of reproducing precisely the evolution of the physical parameters of a nuclear reactor such as the temperature and primary circuit pressure depending on the availability of the different cooling means, in a situation ranging from normal to accidental. These codes are deterministic, and they need to be given the precise dates of the failures whose effects we wish to evaluate. They require substantial calculation resources. There are also other specialized codes for analyzing the complete development of accidents, less precise than CATHARE but faster, such as MELCOR (developed in the USA by *Sandia National Laboratories*) and MAAP4 (developed in the USA by the firm Fauske & Associates, Inc.). Whatever the code used, it is not possible to multiply the simulations infinitely and the expertise required to construct event trees consists of determining a limited number of calculation cases that will make it possible to group together evolutions with the same result (core meltdown or not) in the event tree branches.

The principle of the software developed to carry out dynamic PSAs consists either of trying to automate the approach described above (different variants can be envisaged, but none so far has emerged as the most favorable), or of building a hybrid stochastic model mixing the deterministic evolutions of continuous physical variables and the random discrete state changes due to failures, a model that is exploited via a Monte Carlo simulation. EDF has developed the Pycatshoo tool [CHR 13] for this type of application. Pycatshoo has been used successfully to evaluate the risk linked to some systems, but it remains for now inapplicable to a full scale PSA. There are two reasons for this:

– On the one hand, to be able to model exhaustively the effects of failures and also of the ageing of materials, we would need multi-physical and multi-scale codes, making it possible to calculate the influence of phenomena that occur at microscopic scale (domain of neutronics) on macroscopic characteristics such as the resistance of the vessel steel. Indeed, such codes are not currently available.

– On the other hand, the probability of core meltdown is very low and evaluating it therefore requires a number of simulations that cannot be carried out, even with the massively parallel machines available today. Research on the acceleration of the Monte Carlo simulation has been underway for many years, but the most recent methods still apply only to cases of hybrid stochastic systems that are more simple than those that would be needed [CHR 19].

Despite these difficulties, studies on cases limited to sub-domains of PSAs are possible. For example, in [RYC 18], the author describes some studies carried out at EDF with MAAP4.

3.10. Software for carrying out PSAs

Preamble: Given the rapidly evolving character of everything touching on software, we do not give references in this section. The reader will instantly find fresh information on the Internet through a simple search on the names of the software cited, perhaps supplemented by the names of its editors.

There are two dominant software on the nuclear PSA market: *RiskSpectrum*™, developed and marketed by Lloyd's Register, and the *Risk & Reliability workstation* of which the most known component is the graphical editor *Cafta,* and which has numerous other tools including several quantifiers. The *R&R workstation* is financed by a contribution system organized by the Electric Power Research Institute (EPRI), which is an organization piloting R&D carried out by sub-contractors on behalf of electricity utilities affiliated to them (EPRI). The Nuclear Regulatory Commission (NRC) in the United States has also financed the development of a specialized tool for PSAs called SAPHIRE, but this occupies a small part of the market and is evolving less quickly than *RiskSpectrum* and *Cafta*. These three software benefit from a long history and have integrated many detailed improvements over the course of time. They have editing functions for both event and fault trees, and calculation functions: frequency or probability for undesirable events at different levels: system, isolated sequence or set of event tree sequences, importance factors, etc. *RiskSpectrum*, already a leader in terms of its share of the market, has just made a significant advance on its competitors in 2020, on the one hand, due to the integration of the I&AB calculation method (see section 3.8) and, on the other hand, due to a strong pairing with the KB3 tool from EDF, which makes it possible to generate fault trees automatically from physical system schemas. The version of KB3 adapted to *RiskSpectrum* is called *RiskSpectrum ModelBuilder* and it is maintained and marketed by Lloyd's Register.

Other, more specialized software, can claim to rival these two heavy weights, but on some functions only. The *Andromeda* software from EDF [FRI 14] is a richer and more powerful graphic editor than its competitors, but for now it is only linked to "simple" calculation engines such as SCRAM (see below). It includes original functions, such as the automatic transformation of an Event sequence diagram (section 3.7.1) into an event tree, and various functions for exploring models seen as simple graphs. Starting from there, the tool can calculate the qualitative importance of components using metrics that do not involve probabilities, but only the model structure [RIF 18, RIF 19]. Conversely, the tools XFTA (free) and SCRAM (Command-line Risk Analysis Multi-tool – free and open source) are pure calculation engines: they only support standard Boolean models that do not include all the refinements available in leader tools (the notion of interchangeable basic events, sets of mutually exclusive events, different common cause failure modelings, etc.). XFTA and SCRAM, each developed by a single author, are not benefitting (at the time this book is published) from the support of a company.

The only relatively recent tool including both an event and fault tree graphic editor and a calculation engine is GRIF workshop, marketed by the SATODEV firm that also ensures its development, financed by Total.

Faced with the variety of software available, one question naturally comes to mind: can they be compared, in terms of performances and results? This was a near-impossible task for large models such as PSAs since each software had its own data format. This finding was the origin of the Open PSA initiative launched by a small group of actors in software editing for PSAs [EPS 10]. This led to the creation of the OPSA-MEF (*Model Exchange Format*) data format [EPS 17] that makes it possible to store all the data from a PSA in an XML file. The choice of XML makes it possible to benefit from advances in very many software capable of reading, verifying, representing in graph form and restructuring these types of data. Recent Boolean calculation software such as SCRAM, GRIF and XFTA have the capacity to import and export data in this format. Market heavy weights store their data in relational databases and they do not communicate their structure to their customers. Two obstacles, one technical and the other commercial have restricted use of imports and exports from these databases in OPSA-MEF format. Technically, the OPSA-MEF format does not allow us to represent the *totality* of the data needed for *Cafta* or *RiskSpectrum.* Adaptations to the format are therefore necessary, and it loses its interoperability. Commercially, historic actors have no interest in allowing their clients to export their model since this would allow them potentially to switch to another tool.

3.11. References

[ALI 17] ALIEE H., BORGONOVO E., GLAß M., TEICH J., "On the Boolean extension of the Birnbaum importance to non-coherent systems", *Reliability Engineering & System Safety*, vol. 160, pp. 191–200, 2017.

[ASM 08] ASME RA-S: 2008(R2019), *Level 1/Large Early Release Frequency Probabilistic Risk Assessment for Nuclear Power Plant Applications*, American Society of Mechanical Engineers, New York, 2008.

[BÄK 18] BÄCKSTRÖM O., BOUISSOU M., GAMBLE R. *et al.*, "Introduction and demonstration of the I&AB quantification method as implemented with risk spectrum PSA", PSAM'14, Los Angeles, 2018.

[BAN 17] BANI-MUSTAFA T., ZENG Z., ZIO E., VASSEUR D., "A framework for multi-hazards risk aggregation considering risk model maturity levels", *Second International Conference on System Reliability and Safety*, 2017.

[BIR 69] BIRNBAUM Z.W., "On the importance of different components in a multicomponent system", *Multivariate Analysis*, Vol. 2, Academic Press, New York, 1969.

[BON 92] BON J.L., BOUISSOU M., "Fiabilité des grands systèmes séquentiels : résultats théoriques et applications dans le cadre du logiciel GSI", *Revue de Statistique Appliquée*, vol. 40, no. 2, pp. 45–54, 1992.

[BOR 01] VAN DER BORST M., SCHOONAKKER H., "An overview of PSA importance measures", *Reliability Engineering & System Safety*, vol. 72, no. 3, pp. 241–245, 2001.

[BOR 03] BORGONOVO E., APOSTOLAKIS, G.E., TARANTOLA S., SALTELLI A., "Comparison of global sensitivity analysis techniques and importance measures in PSA, Reliability Engineering & System Safety", *Reliability Engineering & System Safety*, vol. 79, no. 2, pp. 175–185, available at: https:// www.sciencedirect.com/science/article/pii/S0951832002002284, 2003.

[BOU 03] BOUISSOU M., BON J.-L., "A new formalism that combines advantages of fault-trees and Markov models: Boolean logic driven Markov processes", *Reliability Engineering & System Safety*, vol. 82, no. 2, pp. 149–163, 2003.

[BOU 05] BOUISSOU M., "Automated dependability analysis of complex systems with the KB3 workbench: the experience of EDF R&D", *The International Conference on Energy and Environment*, CIEM 2005, Bucharest, 2005.

[BOU 08a] BOUISSOU M., *Gestion de la complexité dans les études quantitatives de sûreté de fonctionnement de systèmes*, Lavoisier, Paris, 2008.

[BOU 08b] BOUISSOU M., "BDMP (Boolean logic Driven Markov Processes) as an alternative to event trees", *ESREL 2008*, Valencia, Spain, 2008.

[BOU 16] BOUISSOU M., HERNU O., "Boolean approximation for calculating the reliability of a very large repairable system with dependencies among components", *ESREL 2016*, Glasgow, 2016.

[BOU 18] BOUISSOU M., "Extensions of the I&AB method for the reliability assessment of the spent fuel pool of EPR", *ESREL 2018*, Hanover, 2018.

[BOU 20a] BOUISSOU M., "Automatic translation from dynamic models to Boolean representation suitable for I&AB quantification", *ESREL 2020*, Venice, 2020.

[BOU 20b] BOUISSOU M., BÄCKSTRÖM O., KRČÁL P., WANG P., "Importance measures in repairable systems", *ESREL 2020*, Venice, 2020.

[CHR 13] CHRAIBI, H., "Dynamic reliability modeling and assessment with PyCATSHOO: application to a test case", *PSAM 2013 – International Topical Meeting on Probabilistic Safety Assessment and Management*, Tokyo, 2013.

[CHR 19] CHRAIBI H., DUTFOY A., GALTIER T., GARNIER J., "Application of the interacting particle system method to piecewise deterministic Markov processes used in reliability", *Chaos: An Interdisciplinary Journal of Nonlinear Science*, vol. 29, no. 6, 2019.

[COL 96] COLLET J., "Some remarks on rare-event approximation", *IEEE Transactions on Reliability*, vol. 45, no. 1, pp. 106–108, 1996.

[EID 16] EID M., ROSATO V., "Critical infrastructure disruption scenarios analyses via simulation", in Setola R., Rosato V., Kyriakides E., *et al.* (eds), *Managing the Complexity of Critical Infrastructures*, Springer, Cham, 2016.

[EPS 05] EPSTEIN S., RAUZY A., "Can we trust PRA?", *Reliability Engineering and System Safety*, vol. 88, pp. 195–205, 2005.

[EPS 10] EPSTEIN S., RAUZY A., HIBTI M. *et al.*, Projet P08-4, "Validation of open-PSA model exchange format for probabilistic safety analyses", Document, IMdR, Gentilly, 2010.

[EPS 17] EPSTEIN S., RAUZY A., HIBTI M. *et al.*, Open-PSA model exchange format, available at: https://open-psa.github.io/mef/_downloads/opsa_mef.pdf, 2017.

[FER 21] FERRANTE F., LEWIS S., PARRY G. *et al.*, "Use of risk insights in the practical implementation of an integrated risk-informed decision-making framework", *Nuclear Technology*, vol. 207, no. 3, pp. 452–459, 2021.

[FRI 14] FRIEDLHUBER T., "Model engineering in a modular PSA", Thesis, LIX, École Polytechnique, 2014.

[HIB 20] HIBTI M., FRIEDLHUBER T., ABDELLATIF T., "Model cartography diving in complex PSA models", *PSA/PSAM*, 2020.

[IAE 92] IAEA, *IAEA Safety Series INSAG-6: Probabilistic Safety Assessment*, IAEA, 1992.

[IAE 99] IAEA, *IAEA-TECDOC-1106 Living Probabilistic Safety Assessment (LPSA)*, ISSN 1011–4289, IAEA, 1999.

[IAE 10] IAEA, *IAEA Safety Standards/Specific Safety Guide SSG-3. Development and Application of Level 1 Probabilistic Safety Assessment for Nuclear Power Plants*, IAEA, 2010.

[KEI 86] KEILSON J., "Stochastic models in reliability theory", *Proceedings of International School of Physic*, North-Holland, 1986.

[KRI 14] KRIAA S., BOUISSOU M., COLIN F. *et al.*, "Safety and security interactions modeling using the BDMP formalism: case study of a pipeline", *Safecomp 2014*, Florence, 2014.

[LAM 75] LAMBERT H.E., "Measures of importance of events", in Barlow R.E., Fussell J.B., Singpurwalla N.D. (eds), *Reliability and Fault Tree Analyses*, SIAM Press, Philadelphia, 1975.

[NKS 21] NKS, "Prolonged available time and safe states", Report, NKS-444, 2021.

[NRC 17] NRC, "Sequoyah integrated deterministic and uncertainty analyses, State-of-the-Art Reactor Consequences Analyses (SOARCA) project", Report, NRC, 2017.

[NUR 85] NUREG, "NUREG/CR-3385 measures of risk importance and their applications", Report, NUREG, 1985.

[NUR 90a] NUREG, "NUREG/CR-4550-Vol.1-Rev.1. Analysis of core damage frequency: internal events methodology", Report, NUREG, 1990.

[NUR 90b] NUREG, "NUREG/CR-4840-procedures for the external event core damage frequency analyses for NUREG-1150", Report, NUREG, 1990.

[NUS 13] NUSBAUMER O., RAUZY A., "Fault tree linking versus event tree linking approaches: a reasoned comparison", *Journal of Risk and Reliability*, vol. 227, no. 3, pp. 315–326, 2013.

[PAG 86] PAGÈS A., GONDRAN M., *System Reliability: Evaluation and Prediction in Engineering*, North Oxford Academic, Oxford, 1986.

[PIL 99] PILLIERE M., MOUTTAPA P., VILLATTE N., RENAULT I., "KB3: computer program for automatic generation of fault trees", *RAMS'99*, Washington, 1999.

[QUI 20] QUINTIN A., DUVAL C., DE DIANOUS V., "Identification des différences de traitement des évènements internes, agressions internes et agressions naturelles extrêmes, lors de l'évaluation du risque d'une installation industrielle", λμ 22, 2020.

[RAU 20] RAUSAND M., BARROS A., HOYLAND A., *System Reliability Theory: Models, Statistical Methods, and Applications*, John Wiley & Sons, Hoboken, New Jersey, 2020.

[RFS 02] RFS 2002-01, *Règle fondamentale de sûreté : développement et utilisation des études probabilistes de sûreté*. Editions Autorité de sûreté nucléaire, Paris, 2002.

[RIF 18] RIFI M., HIBTI M., KANAWATI R., "A complex network analysis approach for risk increase factor prediction in nuclear power plants", *International Conference on Complexity, Future Information Systems and Risk*, SciTePress, pp. 23–30, 2018.

[RIF 19] RIFI M., Modélisation et analyse des Réseaux Complexes : application à la sûreté nucléaire, PhD Thesis, Université Sorbonne Paris Nord, 2019.

[RYC 18] RYCHKOV V., "EDF experience in integrated deterministic probabilistic safety analysis for risk assessment", *Advanced Concepts in Nuclear Energy Risk Assessment and Management*, World Scientific, vol. 1, pp. 245–256, 2018.

[SEV 12] SEVESO, SEVESO 3: Directive 2012/18/UE du Parlement européen et du Conseil du 4 juillet 2012 concernant la maîtrise des dangers liés aux accidents majeurs impliquant des substances dangereuses. Document, SEVESO, 2012.

[SOL 83] SOLOVIEV A.D., "Méthodes analytiques", in Gnedenko B. (ed.), *Voprosy matematicheskoy teorii nadezhnosti (вопросы математической теории надежности)* (in Russian), 1983.

[SWA 99] SWAMINATHAN S., SMIDTS C., "The Event Sequence Diagram framework for dynamic, Probabilistic Risk Assessment", *Reliability Engineering & System Safety*, vol. 63, no. 1, pp. 73–90, available at: https://www.sciencedirect.com/science/article/pii/S0951832098000271, 1999.

[VIL 88] VILLEMEUR A., *Sûreté de fonctionnement des systèmes industriels : fiabilité, facteurs humains, informatisation*, Eyrolles, Paris, 1988.

[ZVO 18] ZVONCEK P., NUSBAUMER O., "Comparison of MCUB and MCS BDD fault tree solution algorithms using Leibstadt nuclear power plant model", *PSAM'14*, Los Angeles, 2018.

4

Structural Reliability: General Presentation, Applications for Nuclear Power Plants

This chapter gives a general presentation of structural reliability analysis (SRA) as a relatively recent scientific discipline, by focusing first on the elementary R-S case, emblematic of SRA and also called the "stress-resistance method", then gives a glimpse, using a general case, of numerical methods for calculating the probability of failure P_f or the reliability index β, both the more classic and the more advanced methods. Then, it succinctly presents different applications of these methods for nuclear power plants, used over 30 years, emphasizing an up-to-the-minute case (optimizing maintenance of a containment vessel). It also mentions the current interest for these methods in other power generation sectors (like hydropower).

4.1. General presentation of SRA

4.1.1. *Why SRA?*

An industrial installation is formed of active components (e.g. pumps, valves, and so on) that modify the installation's operating parameters (flow, temperature, etc.) and passive components, such as structures. Compared to active components, structures are generally less complex (with a smaller number of components and interactions) and therefore house a more limited number of degradation mechanisms and failure modes. These mechanisms follow generally progressive kinetics.

Chapter written by Emmanuel ARDILLON.

The consequences of structural failures can be serious, in terms of risk for society, risk for individuals and risk for the economy and the environment. This is particularly the case in risk-sensitive industries such as nuclear power production. This is why structures are subject to demanding reliability targets.

Activity	Rate $\tau(10^8/h)$	N_e	Rate τ_a $(10^4/yr)$
Mountain Climbing, international level	2,700	100	27
Swimming	350	50	1.75
Smoking	250	400	10
Trawler and deep sea fishing, period 1958–1972	59	2,900	17
Flying (crew)	120	1,000	12
Coal mining	21	1,600	3.3
Automotive transport	56	400	2.2
Construction accidents	7.7	2,200	1.7
Flying (passengers)	120	20	0.24
Home accidents	2.1	5,500	1,1
Work-place accidents	2	2,000	0.4
House or building fires	0.15	5,500	0.08
Structural failure	0.002	5,500	0.001

Table 4.1. *Mortality rate per activity*

In Table 4.1, taken from [MAD 06, LEM 09] with some modifications, τ designates the mortality rate per hour of exposure for 10^8 exposed individuals, N_e the number of hours of exposure per year, and τ_a the annual mortality for 10^4 exposed individuals ($\tau_a = \tau.N_e/10^4$).

Firstly, while the consequence is similar (loss of human life), we see great disparities between the risks of mortality linked to different human activities: this illustrates the differences in the subject of risk perception linked to these activities. Depending on whether exposure to risk is voluntary or not, the risk will be more or less accepted: so, mountaineers passionate about their nonetheless dangerous sport will more easily accept the corresponding risk. Similarly, risks linked to more worrying phenomena will be less well accepted: air transport, which leaves little chance of survival in case of accident, results in a lower annual mortality rate τ_a than car transport, although it is more dangerous.

Second, we see that the risk of mortality linked to structural failures is very low compared to that from other human activities. This practical reliability is the result of a global process involving each stage of the structural life (design, manufacture, installation, commissioning, operation and maintenance, decommissioning). As Figure 4.1 shows [LEM 09], we nonetheless observe relatively significant differences between structures depending on their sector of activity.

Despite a generally high level of reliability, a residual risk of failure remains, as shown by recent events and the emotions they can evoke. We recall, for example, the collapse of the highway viaduct at Genoa (Morandi Bridge) on August 14, 2018, which caused the death of 43 individuals and had an impact far beyond Italy, demonstrating, if necessary, the level of society's demands on the subject of risk. And these demands apply first and foremost to risk-sensitive industries, such as nuclear power production.

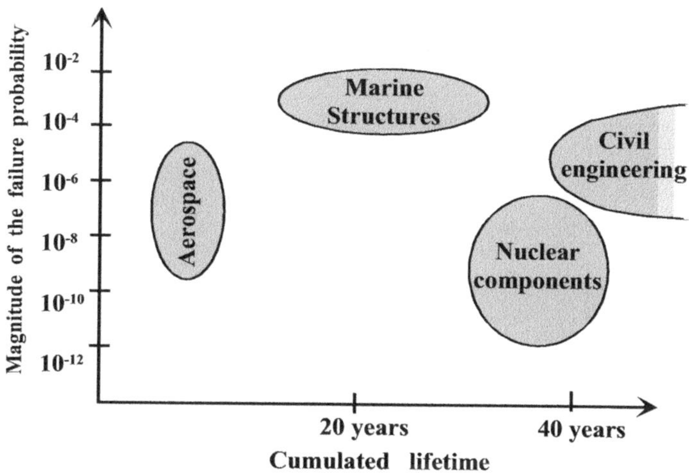

Figure 4.1. *Estimated structural failure probability levels in different branches of industry. For a color version of this figure, see www.iste.co.uk/lannoy/nuclear.zip*

These industries, also subject to performance demands linked to economic functioning in a competitive world, therefore find themselves caught between two demands on which it is necessary or desirable to make a quantitative compromise. It is to this need that SRAs respond, which makes it possible precisely to quantify the risks (or more precisely the occurrence probabilities of feared events). Moreover, it will be noted that the two demands of performance and safety are not opposing demands: a safer installation is also more available for production.

It will be seen that these high levels of reliability, or low probabilities of failure, have implications in terms of numerical calculation methods.

4.1.2. *What does SRA consist of?*

Structural reliability can be defined as the probabilistic approach to structural failure.

4.1.2.1. *SRA: a particular case for treating uncertainty*

SRAs belong to the family of uncertainty treatment approaches, and are both the origin of and a specific case for a methodological reference that is already widely accepted [ROC 08]. Figure 4.2 describes the main stages (simplified version of the source figure).

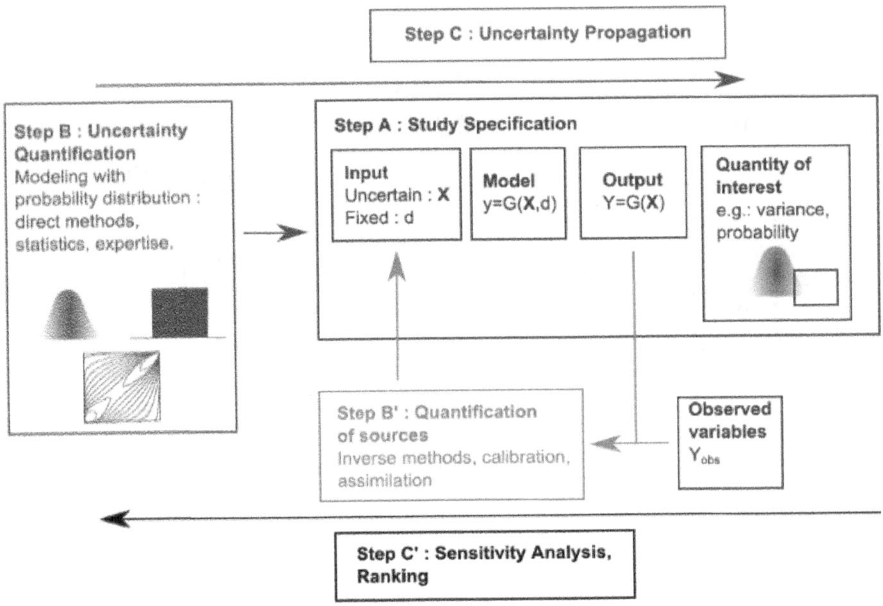

Figure 4.2. *The main stages of the methodology for handling uncertainty. For a color version of this figure, see www.iste.co.uk/lannoy/nuclear.zip*

These stages, also valid for an SRA, are as follows:

– Stage A: problem specification

In particular, this involves defining the model of the structural physical behavior at failure (implemented via the numerical code G, which in SRA is the limit state function), the model's input variables (which characterize the state of the structure), those which are uncertain, the failure criterion (limit state crossing) and the quantity of interest (in SRA, this is the probability of failure or the reliability index β).

– Stage B: quantification of uncertainties

This means identifying the joint distribution of vector X of the uncertain variables, that is, the marginal distributions and their potential correlations.

– Stage C: uncertainty propagation

We propagate the uncertainties affecting the uncertain input variables through the model G to obtain the failure probability studied.

– Stage C': creating a hierarchy of uncertainties, sensitivity analysis

We identify the variables that most influence the probability of failure.

Stages A and B are necessary, stage C forms the core of the approach and stage C' is desirable, often concomitantly with stage C, although not always fundamental.

4.1.2.2. *SRA: one of three families of reliability approaches for components*

As indicated in Chapters 1 (table 1.3) and 5, SRA forms one of four families of reliability approaches for components. Compared to "classic" frequency reliability theory, to be used in the case of operation feedback from significant failure, or Bayesian reliability, to be used to combine operation feedback from potentially limited failure with expert advice or tests, structural reliability is adapted to handling cases where operation feedback from failure is absent (or very limited) and where a physical-mathematical model of structural behavior at the failure is available.

4.1.3. *Old foundations but a recent history*

Evidently, structural reliability relates both to probability calculation[1], a scientific discipline going back to the 17th century, but also to the field of construction, the origin of which is still older and merges with the history of humanity, and to structural mechanics[2], a scientific field formalized since Galileo. Intuitive notions of reliability have been present since the Roman era, which saw the first reliability specialist in the figure of the architect Vitruvius.

Nevertheless, it was not until the 20th century that the happy union of these two parent fields occurred, thanks to more developed conceptual methods and a gradual

1 http:// https://en.wikipedia.org/wiki/Probability#History.

2 https://en.wikipedia.org/wiki/Structural_mechanics.

acceptance of uncertainty. Before the Second World War, only a few pioneers attempted to promote the concepts of uncertainty in structural mechanics [MAD 06], the first being Mayer in Germany, who since 1926, suggested using mean values and the variances of variables in design [MAY 26]. It was left to R. Lévi in France [LÉV 49], and especially in 1947 and to A. M. Freudenthal in the United States to find the bases of current scientific debate [FRE 47].

4.1.4. *SRA: from the R-S elementary case (resistance-stress method) to the general case*

4.1.4.1. *The R-S case (resistance-stress method)*

The most elementary approach consists of modeling a structure by its resistance, written R and its environment by a stress (demand) term representing the applied stress, written S. R and S are generally random, since they are subject to natural variability. Structural failure occurs when the stress (demand) term S exceeds the resistance R: $R - S < 0$. This criterion allows for introduction of the function $G(R, S) = R - S$, which partitions the two-dimensional space (R, S) into a failure domain, characterized by $R - S \leq 0$, and a safety domain, defined by $R - S > 0$. The surface (on the right) separating the two domains, with the equation $R - S = 0$, is called the limit state surface, or the failure surface. The function G is called a performance function, or limit state function, or failure function.

In a dependability context, this approach is known as the resistance-stress method.

The probability of failure P_f is equal to $P(R - S \leq 0)$. It is equal to the integral of the joint density $f_{R,S}$ of the random vector (R, S) on the failure domain, which is:

$$P_f = P(R - S \leq 0) = \iint_{r \leq s} f_{R,S}(r, s)dr.ds \qquad [4.1]$$

When variables R and S are independent, the joint density is equal to the product of the marginal densities, and the result is that the calculation is reduced to a simple integral:

$$P_f = P(R - S \leq 0) = \int_{-\infty}^{+\infty} F_R(y).f_S(y)dy = \int_{-\infty}^{+\infty} f_R(y)(1 - F_S(y))dy \qquad [4.2]$$

where f_X and F_X designate, respectively, the density and the cumulative distribution function (CDF) of the variable X.

These calculations are easy, and are solved by classic 1D or 2D integration methods.

When, as well as being independent, the variables are Gaussian, we directly have an analytical expression of the probability P_f:

$$P_f = P(R - S \leq 0) = \Phi(-\beta), \text{ with } \beta = \frac{\mu_R - \mu_S}{\sqrt{(\sigma_R^2 + \sigma_S^2)}} \qquad [4.3]$$

where β designates the reliability index, μ_X and σ_X designate, respectively, the expectation and the standard deviation of the variable X and Φ the cumulative distribution function of the standard normal distribution. The reliability index used here is Hasofer-Lind's (see FORM, First Order Reliability Method, described in section 4.1.5.2).

It is also possible with this case to evaluate analytically the FORM importance factors α_R^2 and α_S^2 and the corresponding sensitivity factors α_R and α_S (sensitivity indices specific to this method, obtained directly with additional calculation cost and showing the effect of the variable on the reliability index). As a reminder, as will be explained at section 4.1.5.2, the sensitivity factors are the co-ordinates (direction cosines) of the unit vector $\vec{\alpha}$ directing the straight line (OU*), where U* designates the most probable failure point calculated using the FORM method in the standard space where all the variables have been reduced to independent, standard normal variables. We have (with the sign convention $\overrightarrow{OU^*} = \beta.\vec{\alpha}$ used in this chapter (see FORM method, section 4.1.5.2)):

$$\alpha_R = \frac{-\sigma_R}{\sqrt{(\sigma_R^2 + \sigma_S^2)}}, \ \alpha_S = \frac{\sigma_S}{\sqrt{(\sigma_R^2 + \sigma_S^2)}} \qquad [4.4]$$

We can therefore observe that, for this basic case, evaluation of the probability of failure P_f poses no particular difficulty.

4.1.4.2. *The general case*

In the general case, which corresponds to nearly all industrial cases, the structural state is characterized by a state vector X of dimension n, where n corresponds to the number of random input variables $X_{i,\ i=1\ldots n}$ of the physical–mathematical model of the structural behavior at failure.

Formally, the probability of failure P_f is then given by the integral:

$$P_f = \int_{G(x_{i=1,\ldots,n}) \leq 0} f_X(x_{i=1,\ldots,n}) dx_1 \ldots dx_n \qquad [4.5]$$

It is only possible to calculate this integral numerically in some particular and essentially academic cases (despite the counterexample shown in section 4.4) such as the R-S case mentioned previously.

Indeed, in many industrial examples, the G function is a numerical code, the potential complexity of which may result:

– from the multi-physical character of the phenomena (code coupling);

– from the complexity of the modeling (Finite Element code);

– from irregularities (nonlinearities, bifurcations, etc.).

A single call to the G function is therefore costly, and, despite constant progress in computational resources (high-performance calculation, distributed calculations, multithreading), it is necessary to limit the number of calls to the G function [HUC 18].

Moreover, some industrial problems, presenting less complexity in the G function (analytical model), can require a large number of probability calculations (numerous configurations, optimization) [ARD 18].

It is therefore necessary to have effective calculation methods limiting the maximum number of costly calls to the G code. A brief overview is given in section 4.1.5.

We note that in the general case, it is generally possible to define the reliability indices as seen for the R-S case. First of all, there is the reliability index FORM, which will be presented in more detail in section 4.1.5.2, which serves to calculate an approximation of the probability P_f through the relationship $P_f = \Phi(-\beta)$, where Φ designates the distribution function of the standard normal distribution. Then, there is the generalized reliability index that, on the contrary, is calculated from the probability P_f (supposed to have been previously estimated) through the same relationship (reversed). In Table 4.2, we give the correspondence between P_f and β.

P_f	10^{-1}	10^{-2}	10^{-3}	10^{-4}	10^{-5}	10^{-6}	10^{-7}	10^{-8}
β	1.28	2.33	3.09	3.72	4.26	4.75	5.20	5.61

Table 4.2. *Correspondence between Pf and β*

4.1.5. *A brief overview of calculation methods*

Some elements of this section are taken from [LEM 18].

Most of these methods are available in existing reliability analysis tools or in the general reference platform for processing uncertainties at EDF, the OpenTURNS [BAU 17] software, presented below.

4.1.5.1. *Monte Carlo Simulation*

The most basic method and the most well known for calculating probability is the Monte Carlo simulation. It is a very general method that requires no regularity hypothesis on the model used (G function).

The Monte Carlo simulation, applied to reliability analysis, consists of drawing a sample at random of N realizations of the $G(X)$ function. All software and even pocket calculators have a function that generates a pseudo-random sample of a variable of given distribution, often called *aleat*, *rnd*, *random*, whose quality must first be guaranteed. For a reliability analysis, it is enough to count the number q of negative realizations and the frequency q/N gives an estimate of the probability of failure.

This heuristic process is justified mathematically by the law of large numbers (weak and strong) and the central limit theorem that makes it possible to quantify the estimator's speed of convergence and to construct confidence intervals.

It is simple to implement, but its effectiveness is very limited. Indeed, looking for failure, which is a rare event (see Table 4.1), supposes a very low frequency and involves a very large sample, hence a high calculation cost if the G function is complex. More precisely, the convergence speed is in $1/\sqrt{N}$ and we can show easily that we need to make $N = 10^{k+2}$ simulations to estimate a failure probability of 10^{-k} with a precision of 10%, often considered acceptable. The number of calls to the G function therefore quickly becomes prohibitive for the weak probabilities generally encountered. It is therefore necessary to have recourse to other, more effective methods. Nevertheless, the Monte Carlo simulation remains as a reference and a basic tool where it is applicable.

4.1.5.2. *The two main families of classical methods*

They are presented in all reference works in the domain, such as [LEM 09], [MAD 06], [MEL 17] or [ELH 13].

4.1.5.2.1. First-/second-order reliability method approximation methods

A specific presentation of the method is found in [ARD 15]. Figure 4.3 gives a representation of the method.

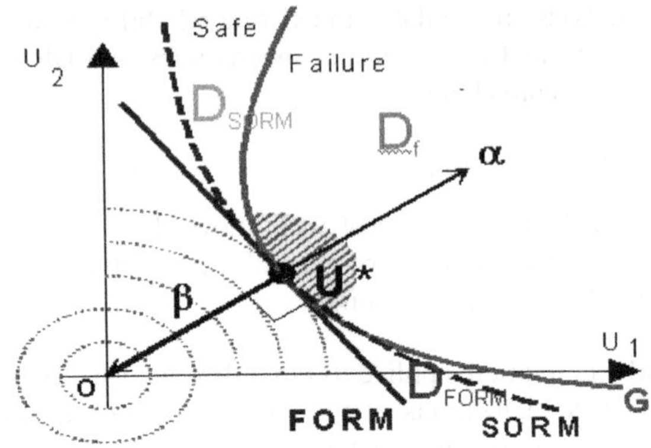

Figure 4.3. *The FORM–SORM method in the standard space. For a color version of this figure, see www.iste.co.uk/lannoy/nuclear.zip*

These methods (first-/second-order reliability method [FORM/SORM]) give an approximation of the probability of failure, which must be controlled systematically. They were developed mainly in the years 1970 to 1990, but developments that enrich them were made until the end of the 2000s. They require three stages:

– First stage: isoprobabilistic transformation

This is a change of random variables, which transforms the initial physical variables into a standard Gaussian space (standard space) where all the variables are scaled to the same scale, and possess interesting mathematical properties: in this space, the main contribution to the probability of failure is situated in the vicinity of the point of the domain closest to the origin, a point written U* or P* depending on the references. The transformation should be isoprobabilistic, so that the failure domain and its transform always have the same probability.

– Second stage: searching the most probable failure point

In this new space, an optimization algorithm in the transformed Gaussian space is used to identify a precise point, the most probable point of failure, also called the design point; the algebraic distance between this design point and the origin of transformed standard space gives the reliability index called Hasofer–Lind, written as β (or β_{HL}). We have: $\overrightarrow{OU^*} = \beta.\vec{\alpha}$, where $\vec{\alpha}$ designates the unit vector directing the straight line (OU*) and oriented from the origin O toward the point U* (see Figure 4.3). The search for the point U* is carried out with the help of an optimization algorithm[3] under equality constraint (it should be on the failure surface). Many

3 https://en.wikipedia.org/wiki/Mathematical_optimization.

algorithms have been developed, distributed into different classes depending on the order of derivation on which they call, such as Hasofer–Lind–Rackwitz–Fiessler's historic first-order algorithm [RAC 79], Cobyla (which does not require calculation of the gradient) or the Abdo–Rackwitz algorithm [ABD 90] (hybrid, second-order algorithm, involving an approximation avoiding calculation of the [Hessian] second-order matrix of the G function at each iteration). Convergence problems may appear, for example if there are several design points (convergence toward a non-global optimum, leaving aside other important optimums) or if the limit state is disrupted (non differentiability, presence of local minima).

– Third stage: approximation

The FORM approximation is of the first order and consists of assimilating the limit state surface in the standard space to its hyperplane tangent at the point U*. From this results an approximation of the probability of failure, estimated by the relationship $P_f = \Phi(-\beta)$, where Φ designates the cumulative distribution function of the standard normal distribution.

The benefit of these methods consists mainly of their substantial effectiveness as well as providing sensitivity factors without additional cost.

Nevertheless, the deviation linked to the probability approximation is not quantified. No estimation of this approximation error is available immediately. Moreover, the search for the most probable failure point U* may not succeed (especially if some regularity conditions of the limit-state surface are not met), or converge toward a local minimum different from the real, most probable failure point (convergence problems).

4.1.5.2.2. Accelerated simulations

Many methods have been developed. The conditional Monte Carlo method always guarantees a reduction of variance, one sometimes spectacular compared to the direct Monte Carlo, but which is not always applicable (it requires being able to isolate one variable from the others when writing the G function) [MEL 17]. The most generally effective method is the importance sampling method. Importance sampling consists of changing the density of samples using a density intended to concentrate the samples in the failure domain, while still conserving enough to draw from the whole space. This new density is called importance or instrumental or proposal density. The bias introduced by the change in density is corrected in the estimator. The effectiveness of this method relies entirely on the choice of the instrumental density. The most efficient choice consists of using knowledge of the most probable failure point U* of FORM. Simulations are therefore centered around this point. Simulations are therefore constrained to fall frequently in the domain of failure and the bias introduced is corrected [LEM 09, chapter VIII]. One alternative possibility (among others) is based on the cross-entropy method [BOU 18].

4.1.5.3. *Some advanced methods*

4.1.5.3.1. Subset simulation

The idea is to reduce the number of Monte Carlo simulations, proceeding by stages in moving, progressively the limit state threshold $G = s_1, s_2 < s_1, \ldots, 0$. We thus define a set of nested sets and at each stage, the probability $P(G(X) < s_i \mid G(X) < s_{i-1})$ is high and can be estimated quickly [AU 01] by means of the Markov-chain generating Metropolis-Hastings algorithm [MET 53].

4.1.5.3.2. Surrogate models

The strategy used here is different: to reduce the unitary cost of a call to the G function, we replace it with an approximate (model) function that is much less costly to calculate. By doing this, a generally unknown approximation error is introduced. Many families of metamodels (surrogate models) have been developed in literature. The basic method is the polynomial response surface method [BAR 11, chapter 7]: the G model is approximated using a polynomial function of the model's (main) input variables. Here, we cite the two other most common families.

4.1.5.3.3. Polynomial chaos

Polynomial chaos is a polynomial stochastic response surface method [BAR 11, chapter 8]: the choice of the polynomial basis depends on the joint density of the random input variables. It consists of breaking down the response on a base of orthonormal polynomials. The stochastic response from a chaos of n variables at order p is developed into a series of $(n + p)!/(n!.p!)$ terms. The coefficients of the chaos give the statistical moments of G and allow calculation of sensitivity factors. Chaos is well adapted to a sensitivity analysis around the average/median and naturally provides the first-order Sobol' indices [SOB 01]. Recall that in sensitivity analysis, Sobol' indices are defined as the proportion of the variance of a random output variable Y, a function of the input variables X_i, due to each of the X_i variables. The finite stochastic elements pair the finite element modeling (FEM) and approximate the uncertainty [GHA 91].

4.1.5.3.4. Kriging

The performance function G is considered the realization of a Gaussian process. An interpolation is made between the calculation points. The essential contribution is the kriging variance that selects the points from the space where interpolation leads to the greatest uncertainty quantifications and so to the selection of new calculation points [ECH 14].

4.1.5.3.5. Classification methods

There is no point seeking to construct G when only its sign is useful. This is the principle of classification methods implemented by support vector machines (SVM) taken from artificial intelligence theories, which construct the limit state surface $G = 0$ [BOU 11].

4.1.5.3.6. Active learning and kriging methods

These recent methods form a synthesis between the last two approaches: surrogate models and classification. They consist of a classification method (failing/non-failing points) based on adaptive kriging. Active learning and kriging (AK) methods make it possible to improve the metamodel in sensitive areas, that is, where the risk of wrongly classifying the points is high. The first method of this family is the Active learning and kriging with Monte Carlo simulations (AK-MCS) method [LEL 17]. This surrogate model technique is used to classify a point basis in the safety or failure domain. To do this, from an initial, moderate-sized design of experiments, an initial kriging metamodel is constructed. A learning function is then evaluated to increase, iteratively, the size of the design of experiments and thus improve the metamodel until it meets a stopping criterion linked to the probability of poorly classifying the points. At each iteration of the method, the probability of failure is evaluated, as in a "classic" reliability analysis using Monte Carlo simulations.

4.1.6. *OpenTURNS: the processing tool for uncertainty quantifications co-developed and used at EDF*

OpenTURNS is a tool for processing uncertainty quantifications developed since 2005 in partnership with EDF-R&D, AIRBUS, PHIMECA Engineering, IMACS and now with ONERA. It is, *a fortiori*, adapted to "structural reliability analyses", studies and covers all stages of the reference framework mentioned in section 4.1.2.1. A detailed presentation of the software is accessible in French [DUT 14] or English [BAU 17]. It covers:

– a C++ library with specific methods and algorithms for processing uncertainty quantifications;

– a textual interface in Python serving as a programming interface (Python module).

It makes it possible to proceed to analyzing the central dispersion, the probability of exceeding a threshold (in particular the probability of failure), the quantiles and the sensitivity.

In the case of complex physical models, it is possible to couple OpenTURNS with an external code (e.g. finite element calculation code).

Two versions a year on average are delivered by the partnership, available in Windows and LINUX.

OpenTURNS is distributed under LGPL license for the IT code and under FDL license for documentation.

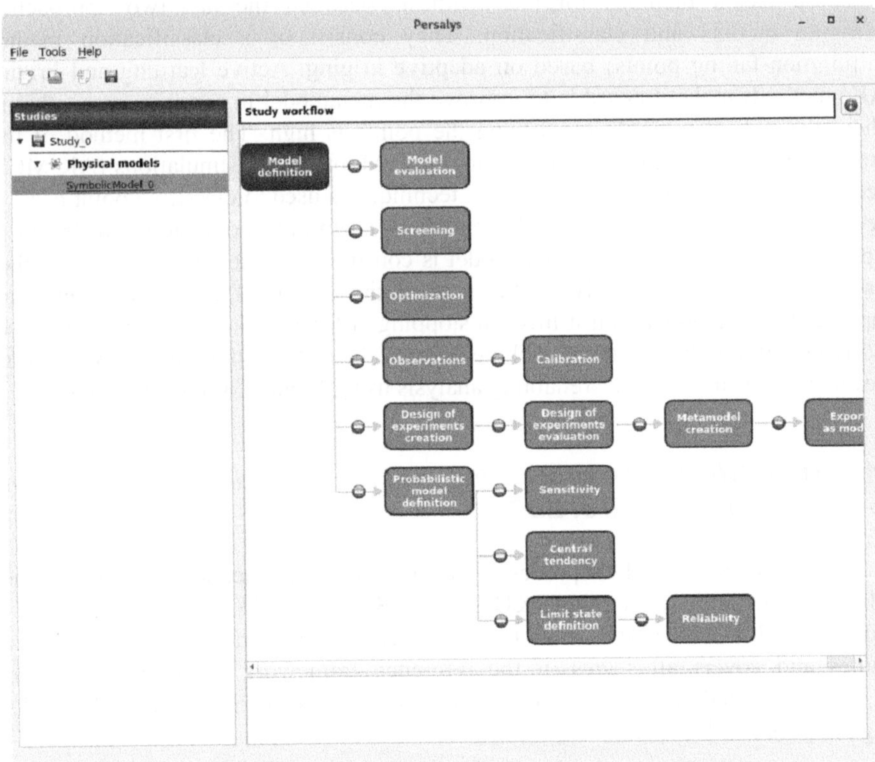

Figure 4.4. *Entry in PERSALYS, OpenTURNS graphical interface. For a color version of this figure, see www.iste.co.uk/lannoy/nuclear.zip*

Recently, a graphical interface intended to manage the code has been developed. It is now called PERSALYS (Version 11 since November 2021 – see Figure 4.4). PERSALYS only gives access to some of the features of OpenTURNS. It gives the useful possibility of automatically running large-scale designs of experiments defined by the user. It is a flexible tool offering a good basis for specific developments linked to different scientific communities.

A broad community of actors has adopted the platform to carry out research studies or prototypes. This broad diffusion enables constant enrichment of the platform using new and ever more effective methods that meet increasingly specific needs.

To conclude, we mention some tools for processing uncertainty quantifications to which OpenTURNS can be compared. First, there is the PROBAN software developed by the Norwegian firm Det Norske Veritas, used at EDF from 1992 to around 2007 essentially for structural reliability studies. We also cite the open source platform URANIE resulting from internal developments at CEA. Less close to home, there is also the general UQLab platform (developed by ETH in Zürich in the MATLAB environment), the commercial package STRUREL (developed by the RCP firm supported by the Technical University of Munich) which, we note, has always been used at times at EDF/HYDRO/DTG for verifying OpenTURNS calculations, the commercial tool COSSAN and the free American software DAKOTA (US DOE, Sandia National Laboratories).

4.2. Structural reliability in the nuclear power generation industry

This section is partly inspired by [LAN 18, chapter 5].

Probabilistic approaches have been used widely in recent years in all industrial sectors, sometimes just to follow rules and regulatory requirements. Following this general trend, the nuclear sector is engaged in probabilistic approaches at the end of the 1980s, most often to complement the regulation or conventional deterministic approach.

Nuclear power plants are reliable in the sense that the overall risk of death arising from their use is much less than the risk posed by other industries or by transport. However, public opinion in developed societies is very sensitive to the risk of accidents in nuclear power plants and more generally to industrial and technological risks. Safety requirements therefore have particular importance for nuclear power plants, and nuclear safety involves many aspects, including the mechanical integrity of systems, structures and components (SSCs). But electricity production is also subject to high demands with regard to competitiveness; these two aspects, which may seem contradictory, necessitate a precise quantification of the risks of failure. Moreover, probabilistic approaches have been used intensively in recent years in all industrial sectors, sometimes simply to follow rules and regulatory requirements. In line with this general trend, EDF has, since the end of the 1980s, entered into the development of probabilistic approaches. The main motivations at EDF in this regard are the better modeling of context and real behavior, better understanding the past and preparing for the future. And the long-term ambition of EDF is to extend the lifetime of its nuclear sites while also improving their performances, especially safety performances, in a competitive environment, hence a growing interest in more precisely estimating margins, in reducing uncertainty quantifications and conservatisms.

Two of the most significant examples in terms of economic issues and research efforts are mentioned below, before we address the case of the pressurizer and above all the case of optimizing the maintenance of vessels subject to a phenomenon of generalized atmospheric corrosion external to their confined environment.

4.2.1. *Optimizing the maintenance policy for steam generators*

The first historical example concerns steam generators (SGs), which are heat exchangers between the primary circuit and its coolant and the secondary circuit, and which produce the steam to drive turbines and electricity-generating alternators. These SGs are formed of a bundle of 3,000–5,000 tubes in Inconel (Nickel-based alloy) with small diameter and thickness, the number of which reflects the desire to optimize the installation's energy yield. With time, these tubes are subject to the various degradation phenomena that lead to cracks, indeed to a risk of fast fracture, the consequence of which is secondary fluid contamination by radioactive water. To ensure the safety of their installation, the operator should obviously prevent this phenomenon, which constrains them to preventatively plug the tubes, presenting a risk of integrity loss that has been judged sufficient. When too many tubes in a SG are plugged, a decrease in the performance of the SG results and consequently a loss of production in the affected plant. Moreover, the cost of replacing a SG is high and this operation should only be considered wisely. Consequently, since the end of the 1980s, an optimization tool for the maintenance policy of the bundle of tubes in SGs, the COMPROMIS code, has been developed [ARD 96]. It is based mainly on a probabilistic fracture mechanics model for the most significant degradation phenomenon (the corrosion of the Inconel alloy and the formation of a crevice at the bottom of the tube). The use of feedback data from SGs to characterize uncertainty quantifications makes it possible to evaluate the tube rupture probabilities appropriate for a given SG, as well as the evolution of the number of tubes to plug, depending on time, and consequently the SG's technical lifetime.

4.2.2. *Risk of fast fracture of PWR reactor pressure vessels*

The second historical example, launched in 1993, is a probabilistic approach to the fast fracture of reactor pressure vessel (RPV) in pressurized water reactors (PWR) nuclear power plants. This risk involves RPV shells in low alloy ferritic steel.

Pressurized water reactor RPVs are subject to a potential residual risk of fast fracture resulting from the three factors:

– First, the possible or proven presence of previously existing planar cracks, due to manufacturing; these cracks are situated in the heat-affected zone (HAZ), located at the boundary between the body of the shell and the internal anti-corrosion clad, the location of which can give rise to significant constraints.

– Second, in some exceptional or accidental operational situations, plants can undergo severe thermo-hydraulic transients of the pressurized cold shock type (e.g. following an injection of cold water destined to cool the reactor, "urgently", for example, in case of a pipe break on the primary circuit). These transients are characterized by a sudden cooling of the water supplying the RPVs (and so of their steel wall) as well as a significant re-pressurization. This results in significant stresses in the RPV wall, in particular at the level of any possible cracks that may already exist, and more precisely the appearance of stress gradients likely to lead to crack opening; moreover, the drop in temperature, if it is pronounced enough during transition, can lead the steel to work in the transition zone between its ductile and brittle behavior domains.

– Finally, a cumulative neutron irradiation that increases with the time spent in operation and which results in an increasing embrittlement of steel of RPV shells situated in the central zone of the reactor, that is, of diminishing the ductility and toughness of the steel. This embrittlement manifests physically in the potential appearance of cleavage (trans-granular rupture), and then fast fracture in these zones. It will be noted that this will be the only one of the three factors that changes over time.

Risk evaluations for steel shell fast fracture consist of propagating the uncertainty quantifications affecting the input parameters of the fracture mechanics model for the various thermohydraulic transients analyzed, retained for their conservative character. These evaluations aim to avoid the uncontrolled accumulation of conservatisms (margins) operated by the deterministic approach on the model's different input parameters. They have given rise to international benchmarks such as [FAI 09] or more recently [DAU 12].

Many other application examples have been handled in the past 25 years. These examples, such as that of the pressurizer mentioned below, show that it is possible to use probabilistic approaches in evaluations linked to safety such as an analysis tool closer to industrial reality, complementing regulatory or conventional deterministic approaches.

4.3. The pressurizer, an example of an exploratory exercise in the application of probabilistic approaches

This application is the subject of a more detailed presentation in [LAN 18, chapter 5].

In the wake of applications on components with high stakes in terms of importance for the safety, economic performance or lifetime of PWR nuclear power plants, an application has been started for illustrative purposes on a component of the main coolant system: the pressurizer.

The pressurizer of PWRs is a safety-significant component located on one hot leg of the reactor coolant system (RCS). Its main role is to keep the primary pressure constant (at around 15.5 MPa) to prevent the RCS water from boiling. It is a replaceable component, not considered critical in terms of lifetime. Its main body is composed of cylindrical welded shell rings. These ferritic steel rings and their welded joints may experience cumulated thermal aging resulting in fast fracture damage.

Despite the very unlikely character of the presence of cracks under the shell clad, taking account of the manufacturing process used, regulation imposes the need to justify the absence of risk of fast fracture of the pressurizer.

The pressurizer's conventional integrity assessment is relative to an underclad crack situated in the shell or in the welded joint, apt to give rise to a fissure. The most severe operating conditions correspond to calculated thermal ageing for the welded joint, and to hydraulic testing of the RCS. We consider the highest phosphorous content of the base material of 900 MWe pressurizers. In the temperature conditions affecting the pressurizers, it is in fact phosphorous that comes to separate from the grain joints of the low alloy ferritic steel and embrittles the steel, thus increasing the residual risk of brittle rupture where there is a crack.

The current integrity evaluation rule is written as: $K_{CP}(C_S.L) \leq K_{IC}$, where K_{CP} represents the stress intensity factor calculated at the bottom of the crack for the load L considered to be multiplied by the regulatory security coefficient C_S valid for the load category L.

This section summarizes the results of an exploratory probabilistic study carried out with different hypotheses on the input parameters of the mechanical-reliability model. For a given set of parameters, the critical configuration corresponds to the moment where the margin factor (MF) reaches its minimum authorized value, given by the security coefficient C_S (as it happens, 1.6 for the hydraulic test).

To identify margins that potentially exist, an exploratory probabilistic approach has been engaged in addition to conventional deterministic evaluation.

First of all, a probabilistic methodology has been developed. Its main benefit is that it provides a solution to define a reference probability that can be used to compare the reliability of the pressurizer for the damage considered (here, the brittle rupture due to thermal ageing). It is similar to the solution adopted in the Eurocodes [CAL 01]: it considers that the probabilities calculated are notional probabilities (i.e. conventional), and that the reference probabilities should be evaluated in a way that is coherent with the way in which the probabilities of failure are evaluated.

Two configurations were then tested, corresponding to (a_{conv} = 20 mm, RT_{NDT} = 74.35°C, t = 40 years, where a designates the height of the conventional crack and RT_{NDT} the ductile to brittle transition temperature of the steel). They correspond to two possible values of the pressure applied during the hydraulic test (HT) of the RCS, that is, P = 22.8 MPa (Configuration no. 1, "full rate" HT) and P = 20.6 MPa (Configuration no. 2, "reduced rate" HT). For each configuration, it is estimated that the temperature of the HT for which the MF F_M reaches exactly the value of the security coefficient: 1.6–40 years. In other words, the conventional crack remains stable for 40 years (at least) for HT temperature higher than the calculated temperatures. We find the following values: (P = 22.8 MPa, T_{TH} = 95.06°C) and (P = 20.6 MPa, T_{TH} = 89.88°C). Considering the demands recalled above, these are realistic values of T_{TH}.

For these two cases, the probability of failure of the conventional crack at 40 years is equal to 0, the same as for the pressurizer which is necessarily lower than or equal to it. This in itself is a significant result of this study. It is not therefore necessary to apply the probabilistic methodology described previously since it is no longer necessary to have a reference probability: a zero probability of failure is necessarily acceptable. This remark would also be valid for very low, non-zero probabilities.

Note that the reference probability resulting from the methodology was also zero: this shows the methodology's strict character. It could be useful to adapt this methodology by envisaging defining a new, non-zero reference probability in such a case.

In the end, application of the probabilistic methodology proposed is not strictly necessary: considering the hypotheses on the boundaries of the two distributions envisaged (toughness and height of the crack), the study amounts to identifying the instant from which the lower boundary of the distribution of the MF becomes lower than 1. The study is finally reduced to a boundary method (interval approach). This instant corresponds to the moment where the probability of failure initiation ceases to be zero.

In addition, the results show that it is possible to envisage extension of the pressurizer's operational lifetime from 9 to 14 years beyond the lifetime initially predicted at design, depending on the hydraulic test rate considered, without any risk of fast fracture. These results are of course linked to the modeling hypotheses of uncertainty quantifications affecting the input data of the mechanical model (especially the lower boundary of the toughness distribution and the height of the conventional crack).

These results show the benefit of the probabilistic approach for identifying existing margins and for complementing the conventional deterministic justification. To complete this exploratory exercise, it may be interesting to refine the modeling of the uncertainty quantifications affecting the input data (e.g. uncertainty in measuring the concentration of phosphorous).

4.4. Probabilistic optimization of the maintenance of nuclear power plant steel components

4.4.1. *Introduction*

Section 4.2 gives a glimpse of EDF's experience in the subject of structural reliability applications. It seemed useful to present more particularly an example of these approaches, by selecting the most recent study, which is therefore the subject of this section, and which is the subject of a detailed presentation in [ARD 22].

Some steel structural components situated in the secondary part of PWR nuclear power stations are subject to atmospheric corrosion in a confined environment, affecting their bottom via the external surface. For one of them (this is a vessel), several campaigns to measure the thickness of the bottom of the vessel are available over the course of time over a population of measurement points. These points are situated in the peripheral area of the bottom where the corrosion is the most active. Measurements show a significant effect of corrosion with an overall thickness loss over time. It is possible to evaluate, from these measurements, a thickness loss kinetics distribution and a residual thickness distribution over time.

This section presents the probabilistic approach implemented as well as its main results, useful for optimizing maintenance on this component.

It is a probabilistic predictive approach making it possible to obtain the evolution of the residual thickness measurements (on which the failure criterion hinges) over time. This is a prognosis type of approach. For each year to come, it is therefore possible to calculate at each of the 23 measurement points:

– a residual thickness distribution, real or measured;

– a risk of thinning points.

It is then possible to combine, at any given instant, these individual risk evaluations to obtain an overall risk of insufficient thickness from the whole population of measurement points at a desired instant, and so to follow the evolution of the overall risk of insufficient thickness over time.

The approach presented therefore relates to SRA. Compared to the deterministic approach, it will be noted that the probabilistic approach has the double advantage of allowing rigorous handling of uncertainty quantifications (rather than a stacking of margins with a conservatism level that is not always well managed), as well as a precise evaluation of the risks.

The presentation retained in this section follows the logical unfolding of the stages of the "uncertainty quantifications" reference presented in section 4.1.2.1.

4.4.2. *Specifying the problem (stage A)*

4.4.2.1. *Failure criterion*

The failure criterion, in line with the component's earthquake resistance, relates to the thickness measured Th_{meas}. For points already repaired, the criterion is $Th_{meas} \leq 0$ mm, for points not repaired, the criterion is $Th_{meas} \leq 2$ mm, where Th_{meas} designates the thickness measured.

4.4.2.2. *Significance of corrosion*

Even if the available data seem to indicate a "visible" corrosion effect, one subject to monitoring by the operator, it is not unhelpful to seek confirmation of this by means of a statistical test.

The Wilcoxon signed-rank test is used to compare paired samples of thickness measurements made during measurement campaigns from 2001, 2010, 2012, and 2016. All comparisons (even those from the campaigns that were closest together) concluded there was a significant corrosion effect.

4.4.2.3. *The selected model*

By defining the overall measurement error written ε, we have at any instant t the relationship [4.6]:

$$Ep_{meas}(t) = Ep_{real}(t) + \varepsilon \qquad [4.6]$$

Note that ε is called the overall measurement error (OME) since, aside from the measurement process error itself (here, ultrasound), it also encapsulates the measurement's random location error. Traditionally, it is considered independent of the real thickness.

It should be noted that in our case, we do not have an appropriate physicochemical model of the type of corrosion involved. It is assumed that the phenomenon is linear and it comes down to knowing an annual corrosion rate, written as $c_{0i,\ real}$ (negative with the convention used). In the absence of a physical model, this rate should be evaluated from the thickness measurements.

If we now wish to predict the thickness measured at a point k at an instant t_j, knowing the thickness measured at t_0 $Ep_{meas,k}(t_0)$, we use the following relation:

$$Ep_{meas,k}\,(t_i) = Ep_{meas,k}\,(t_0) - \varepsilon_0 + c_{0i,\ real} \cdot (t_i - t_0) + \varepsilon_i \qquad [4.7]$$

which is written as:

$$Ep_{meas,k}\,(t_i) = Ep_{meas,k}\,(t_0) + c_{0i,\ real} \cdot (t_i - t_0) + \varepsilon' \qquad [4.8]$$

where ε' follows a normal distribution with a zero mean and a standard deviation equal to $\sigma_\varepsilon.\sqrt{2}$.

These relations can also serve to estimate the real corrosion kinetics $c_{ij,\ real}$ between instants t_i and t_j. There results:

$$c_{ij,\ meas} = c_{ij,\ real} + (\varepsilon_j - \varepsilon_i) / (t_j - t_i) \qquad [4.9]$$

ε_i and ε_j designate, respectively, the overall measurement error made at times t_i and t_j ($t_j > t_i$). These are two independent random and identically distributed variables (i.i.d). $c_{ij,\ meas}$ designates the annual rate of evolution of the thickness measured between times (years) t_i and t_j. $c_{ij,\ real}$ designates the annual evolution kinetics of the real thickness between times (years) t_i and t_j.

It finally reduces therefore, for each point of measurement, to a problem with two independent random variables (the state vector X = ($c_{0i,\,real}$, ε')). And the limit state function G of the problem (for an individual measurement point) is as follows:

$$G(X) = G(c_{0i,\,real}, \varepsilon') = Ep_{meas,k}(t_0) + c_{0i,\,real} \cdot (t_i - t_0) + \varepsilon' - \text{threshold} \qquad [4.10]$$

This therefore brings us to the basic R-S case mentioned above (stress-resistance method, section 4.1.4.1), posing for example:

$$R = Ep_{meas,k}(t_0) + \varepsilon' - \text{threshold} \qquad [4.11]$$

$$S = -c_{0i,\,real} \cdot (t_i - t_0) \qquad [4.12]$$

R illustrates the contribution of the OME, S that of the corrosion kinetics.

For the set of measurement points N, this is a problem with 2.N independent random input variables (2.N dimension), which reduces to N problems of two-dimension.

4.4.3. *Uncertainty quantification (stage B)*

This involves identifying the joint distribution of two random input variables. Since these two variables are independent, this consists only of identifying their marginal distribution.

4.4.3.1. *Variable ε'*

It has a zero mean. This amounts to identifying the marginal distribution of ε. Classically, for a measurement error we take normal distribution. Its standard deviation remains to be evaluated. This is done on expert advice, and also in coherence with the measurement data. Values $\sigma_\varepsilon = 0.2$ or $\sigma_\varepsilon = 0.3$ are considered.

4.4.3.2. *Variable $c_{0i,\,real}$*

According to relation [4.9], we estimate first the distribution of the evolution rate of the thickness measured $c_{0i,\,meas}$ according to the available data, then the distribution of the real kinetic is deduced from this.

The first question is the choice of measurement campaigns to be retained for evaluating this rate. To minimize sensitivity to the OME, we choose the campaigns the most distant in time (2001 and 2016).

The first analysis with the Shapiro–Wilk normality test [AGO 86] shows that the normality hypothesis is (very) acceptable for this distribution. It is then completed by an analysis of goodness-of-fit to the data from the available distributions by means of goodness-of-fit tests such as Bayesian information criterion, a visual analysis (by Q-Q plot) and considering some constraints specific to this example (such as theoretically infinite support of the distribution) [ARD 21]. It appears that no distribution gives better results than normal distribution on all criteria. Normal distribution can therefore be retained to model $c_{0i,\ meas}$.

Since the two variables ε' and $c_{0i,\ mes}$ are Gaussian, it then results via relation [4.9] that $c_{0i,\ real}$ is also a Gaussian variable [ARD 21].

4.4.4. *Uncertainty propagation: calculating the overall risk of thinning points (stage C)*

This calculation is made in two stages:

– calculation of individual underthickness risk written as $P_{def}(t_i, k)$, at measurement point k and at instant t_i;

– calculation of the overall underthickness risk on the population of N measurement points cumulating the individual risks using the formula:

$$RG(t_i) = 1 - \prod_{k=1}^{N}(1 - P_{def}(t_i, k)) \quad [4.13]$$

This is the probability, at instant t_i, that there is at least 1 point among the N measurement points that is insufficiently thick.

For the first stage, we clearly have:

$$P_{def}(t_i, k) = P[Ep_{meas,k}(t_i) < \text{threshold}] = P(G(c_{0i,\ real}, \varepsilon') < 0) \quad [4.14]$$

As we have seen, we have returned to the "Gaussian R-S" case mentioned in section 4.1.4.1, and we evaluate $P_{def}(t_i, k)$ analytically in the following fashion:

$P_{def}(t_i, k) = \Phi(-\beta)$, with

$$\beta = \frac{Ep_{meas,k}(t_0) - \text{threshold} - \mu_{0i,real} \cdot (t_i - t_0)}{\sqrt{(2.\sigma_\varepsilon^2 + \sigma_{0i,real}^2 \cdot (t_i - t_0)^2)}} \quad [4.15]$$

It will be noted that the FORM sensitivity factors can easily be obtained (see section 4.1.4.1) for each individual risk [ARD 21].

Finally, the overall underthickness risk follows the evolution given in Figure 4.5.

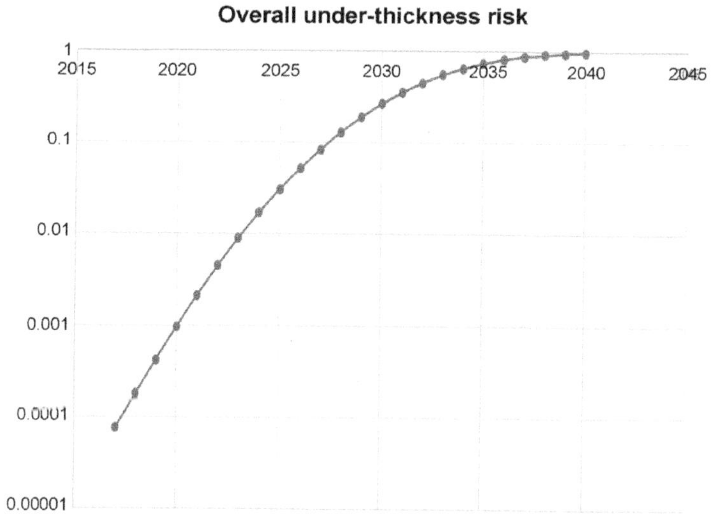

Figure 4.5. *Overall under-thickness risk as a function of time (log scale). For a color version of this figure, see www.iste.co.uk/lannoy/nuclear.zip*

4.4.5. *Using probabilistic results: determining points to repair*

4.4.5.1. *The chosen approach*

Here, we describe the way in which overall risk evaluations resulting from the previous stages of the approach are used.

The criterion retained is the overall risk of insufficient thickness in 2033, $RG(t_i = 2033)$.

If $RG(t_i = 2033)$ exceeds 10%, we will seek the minimal set of n_p measurement points to repair (patch). This makes it possible to minimize the repairs to be made. This set corresponds to the points of greatest individual risk, classified by decreasing risk, keeping only the first points, making it possible to reach an overall risk less than or equal to 10. In other words, we then select n_p points such that:

$$RG_{n\backslash np}(t_j = 2033) < 10\% \text{ (and } RG_{n\backslash np+1}(t_j = 2033) \geq 10\%) \qquad [4.16]$$

In a second phase of the study, the level of risk tolerated has been taken equal to 5%. This also makes it possible to take account of the modeling hypotheses and especially of the fact that, at stage B, quantifying uncertainty, the samples used for fitting the distribution of thickness loss kinetics are small.

4.4.5.2. *Numerical results*

The number of points to repair depends on the overall risk criterion chosen. As seen previously, it was first fixed at 10%. We are therefore going to select the points to repair sequentially, starting from the point of greatest individual risk and dealing them in line with their contribution to the overall risk until we arrive at a risk below 10%. For those points not repaired, these will of course be the thickness points that measured lowest in 2016.

In total, the points to repair before 2033 are:

– one point already repaired in 2016;

– seven points not repaired in 2016.

By repairing them, we arrive at an overall risk equal to only 8.5%.

The average number of thinning points after removing the repaired points would then be 0.089 in 2033, which is very low.

If we wish to reduce the risk and to tolerate a risk of thinning of only 5%, then we need to repair two more points; the overall risk is then 5.03%. If we wish to be strictly below 5%, we need to repair a third point and the overall risk is then 4.04%.

In all cases, it seems that it is not necessary to replace the bottom of the vessel, it is enough to repair a number of the points measured (8 to 11).

4.4.6. Conclusion and perspectives on this application

4.4.6.1. *Industrial benefit*

First of all, compared to the deterministic approach, it appeared that the probabilistic approach has the double advantage of permitting a rigorous treatment of uncertainty quantifications (rather than a stacking of margins with a conservatism that is not always well managed), as well as a precise evaluation of the risks.

Considering the year 2033 as the temporal horizon, we find that without maintenance intervention, the overall risk in 2033 is comprised between 54.8% ($\sigma_\varepsilon = 0.3$ mm) and 58.6% ($\sigma_\varepsilon = 0.2$ mm). The variability of the measurement error therefore has little impact on the result.

Moreover, it also appears that without the same intervention, the mean number of underthickness points will still remain less than 1 in 2033: we find $N^{av}_{th} = 0.76$. This number may seem low.

Finally, if we accept an overall underthickness risk of 10% in 2033 (decision criteria), it is necessary to repair eight points out of 23, which are the eight of greatest individual risk: seven are new points to repair and one is a point already repaired. If we retain a risk of 5%, we need to repair two or three additional points.

In all cases, the study has shown that a repair to the bottom of the vessel is needed to meet the failure criteria.

In contrast, it appears that it is not necessary to proceed to replacing the bottom completely, and local repairs (made by applying patches) on a number of points resulting from the study make it possible to guarantee the state of the vessel studied, which contributes to a noticeable optimization of preventive maintenance of this equipment.

These results are in the process of being validated internally. They will be accompanied by additional calculations (sensitivity studies). As additional measurements should be available in the coming years, the possibility to update the thickness loss distribution by treating the non-normality case of the thickness loss is also envisaged. The recommended maintenance interventions should take place in the coming years.

4.4.6.2. *Scientific interest*

First of all, structural reliability will have confirmed through this study its aptitude for solving an industrial problem.

This study will also have made it possible to identify an industrial problem that can be solved with the most basic structural reliability approach, that is, the R-S case (stress-resistance method). Of course, this is not the general rule where physical modelings tend to become ever more complex to better correspond to reality, but this reflects the situation of this problem in terms of physical and regulatory modeling. The R-S case is not therefore a merely academic example.

Despite its simplicity, this problem will have been the occasion to tackle, in the perhaps temporary absence of an adequate physical degradation model, a development phase of the most relevant model considering the information available. This is an element of originality of this study, since the physical model (behavior, degradation) is often a piece of input data in SRAs. Doing this, it will have led to an analysis of the significance (highly probable) of the degradation envisaged, a question that is not usually asked in studies, where the degradation model is already specified.

The study will also have confirmed the need to be able to evaluate an overall risk (union of events) and so handle system events. This point has recently (2019) been the subject of integration with OpenTURNS, which since version 1.14 has made it possible to calculate the probabilities of system events. It will therefore have confirmed the relevance of this evolution of the software. It will also have shown the need to have sensitivity indices for these system event probabilities.

4.5. Structural reliability for hydroelectricity – the reliability of penstocks: evaluation of calculation values for mechanical strength diagnostics

Before concluding this presentation of applications of structural reliability to problems in nuclear power stations, including some methodological aspects, it seems useful to show that these methods, far from being limited to the peripheral area of nuclear power stations, are applied in other domains linked to power generation. In this section, we give an overview of an application in the hydropower domain: reliability studies on penstocks.

This study [ARD 18, BRY 20] consists of an industrial application of margin calibration methods, here relating to characteristic values and not to partial coefficients. In the hydropower sector, penstocks are pipes ensuring the flow of water under between water sources and the turbine. These are thin pipes with a diameter to thickness ratio generally comprised between 25 and 250. The diameter of a penstock is comprised between 0.4 m and 5.5 m. The nominal thickness of the pipes, defined at design, depends on the maximal operating pressure; it is comprised between 5 mm and 50 mm.

Over time, the margin of a penstock in operation regarding the risk of plastic collapse is likely to decrease, especially after thinning caused by corrosion. Diagnoses are regularly carried out to check the mechanical integrity of penstocks faced with this risk. They rely on calculating a deterministic MF that depends on conservative design values (characteristic values) intended to cover the uncertainty quantifications affecting some of the model's input variables (ultimate tensile strength R_m, yield stress R_e and loss of thickness Δe). This MF is the ratio of the allowable stress considered written f to the maximum allowable operating stress, written as σ_c: $MF = f/\sigma_c$. f is expressed as a function of the calculation values of R_e and R_m; most frequently: $f(R_e^d; R_m^d) = \min\left(\frac{R_e^d}{1.5}; \frac{R_m^d}{2.4}\right)$.

The calculation values R_m^d (respectively, R_e^d) and Δe^d correspond to quantiles of R_m (respectively, R_e) and Δe shifted by a number η_{R_m} (respectively, η_{Δ_e}) of standard deviations from the mean value. These coefficients η are called standard deviation multipliers. They are often taken equal to 2 in diagnoses, considering the different uncertainty quantifications affecting these variables.

The question is to find out whether these conventional coefficients η can be chosen in a rational way: to what level of risk do they correspond? What should be their value to respect a given reliability target level β^{T} taking account of the coefficients applied to define the allowable stress?

To answer this question, a probabilistic approach was performed. It relies on a mechanical reliability model that conforms to the analysis criteria of standard [BS 15] for underthicknesses situated in parent metal (excluding welds).

This model includes three independent random variables in its input: R_m, Δe (variation in thickness resulting from the initial manufacture, then to corrosion) and ε conveying the uncertainty quantifications on the linear correlation obtained between R_e and R_m: $R_e = A\,R_m - B + \varepsilon$. Some distributions have been deduced from measurements taken from a representative panel of diagnoses on operational penstocks. Failure occurs when the stress applied σ_c exceeds the plastic yield stress $\sigma_f\colon = \sigma_f - \sigma_c \leq 0$,

with $\sigma_f = \min\left((R_e + R_m)/2; 0.85 \cdot R_m\right)$

and $\sigma_c = f/FM.\left(e_{\text{design}} + \Delta e^d\right)/\left(e_{\text{design}} + \Delta e\right)$.

The annual probability of failure during the year N is the intersection of events $\mathrm{G}(N-1, \Delta e) > 0$ and $\mathrm{G}(N, \Delta e + \Delta e_N) < 0$. Considering the decrease of the function G over time, the annual probability of failure is $\mathrm{P}(N) = \mathrm{P}(G(N-1, \Delta e).\,G(N, \Delta e + \Delta e_N) < 0)$. It is evaluated using the OpenTURNS Python (and C++) library and the dedicated PERSALYS-Penstock tool (Figure 4.4) of which OpenTURNS is the calculation engine, mainly using the FORM method (adapted to process an intersection of events) combined with importance sampling.

The target reliability levels β^T are retained following the recommendations of standard [BS 15] for non-redundant components whose failure leads to “severe” to “very severe” consequences. To cover all the possible configurations of penstocks, an analysis grid has been defined, with a combinational approach as complete as possible.

This study has finally made it possible to show that taking standard deviation multipliers η equal to 2 leads, in most cases, to conservative results. With this choice of multipliers, an MF equal to 1 may correspond to a probability of failure through plastic yield of the penstock that is less by several tens to the target probability allowed for this type of work. It would therefore be possible to take calculation values for the material and the residual thickness that are closer to the average value ($\eta < 2$). More precisely, we see in the retained analysis grid that, provided the standard deviation of the thickness loss does not exceed 0.75 mm with a nominal thickness of at least 8 mm, multipliers taken equal to 1 lead to an annual probability of failure that respects the target level. These cases correspond to a variation coefficient of the residual thickness less than or equal to 0.1. Only some particular configurations, linking very high dispersions of thickness losses (standard deviation of the thickness loss above 0.75 mm) and pipes of low nominal thickness (5 mm), would require higher η multipliers to be applied.

4.6. Conclusion

This chapter has made it possible, on the basis of more than 30 years' experience in Structural Reliability, to demonstrate the diversity of applications addressed in terms of:

– the physical degradation or failure phenomena at work;

– the challenges (operation time, predictive maintenance extension of lifetime, under safety constraints);

– the methods available or applied.

It also has shown that the application perimeter of structural reliability is not limited to nuclear power generation, but also concerns many other domains.

Applications have shown that SRA is a pluridisciplinary approach at the cross-roads of multiple diverse competencies: probabilities, structural mechanics, statistics, materials environment, operation and maintenance, risk analysis, etc. They will also have shown that the successful implementation of SRA relies especially on direct collaboration between the operator and the researcher, essential for understanding, analyzing and interpreting. We note on this topic the importance of the data making it possible to quantify the uncertainties relating to the input data of models: materials (resistance), loads, potential defaults, geometry, etc.

At the time this chapter is written, perspectives of development for these approaches seem to on offer in all domains of power generation, including wind power (offshore in particular), which will be the occasion, in the context of the European HIPERWIND project launched in December 2020, to take advantage of the advanced probabilistic calculation methods mentioned in this chapter (as it happens, AK methods, applied to evaluation of fatigue damage then to fatigue reliability analysis).

Although the methods are sometimes in advance of application needs encountered in the nuclear domain (e.g. time-dependent reliability), applications are also at the root of methodological development needs or of the evolution tools for processing uncertainty quantifications used at EDF (i.e. the OpenTURNS software and its graphical PERSALYS interface), such as the system event processing (overall risk calculations or annual failure probability calculations), improvement of existing methods and algorithms to compute system event failure probabilities, adaptation of methods to some particular cases (numerical integration, conditional Monte Carlo), sensitivity indices, especially for system events, robustness of reliability results *vis-à-vis* epistemic uncertainty quantifications, to cite only a few current points of interest. It should be noted that, as an illustration, evaluations of the robustness of failure probabilities with respect to epistemic uncertainties have also been carried out on an exploratory basis using the info-gap method [BEN 06, AJE 20].

4.7. References

[ABD 90] ABDO T., RACKWITZ R., "A new beta-point algorithm for large time-invariant and time-variant reliability problems", *Reliability and Optimization of Structures*, pp. 1–11, 3rd W.G. 7.5, IFIP Conference, Berkeley, USA, 1990.

[AGO 86] d'AGOSTINO R.B., STEPHENS M.A., *Goodness-of-fit Techniques*, CRC Press, Taylor & Francis Group, Boca Raton, USA, 1986.

[AJE 20] AJENJO A., ARDILLON E., CHABRIDON V. *et al.*, "Robustness evaluation of reliability assessments of penstocks using info-gap method", *e-Proceedings of the 30th European Safety and Reliability Conference and 15th Probabilistic Safety Assessment and Management Conference*, Venice, Italy, 2020.

[AJE 22] AJENJO A., ARDILLON E., CHABRIDON V. *et al.*, "An info-gap framework for robustness assessment of epistemic uncertainty models in hybrid structural reliabiliy analysis", *Structural Safety*, vol. 96, 2022.

[ARD 96] ARDILLON E., PITNER P., LAPEYRE B., "Implementation of variance reduction techniques using Monte Carlo stratified sampling in the COMPROMIS Code Version 2", *Proceedings of the 7th Specialty Conference on Probabilistic Mechanics & Structural Reliability*, pp. 692–695, ASCE, Reston, VA, USA, 1996.

[ARD 97] ARDILLON E., VILLAIN B., BOUCHACOURT M., "Probabilistic analysis of flow-accelerated corrosion in French PWR: the probabilistic module of BRT-CICERO version 2", *Proceedings of ICOSSAR n°7 (International Conference on Structural Safety and Reliability)*, Kyoto, Japan, 1997.

[ARD 03] ARDILLON E., BARTHELET B., MEISTER E. *et al.*, "Reliability-based approaches for safety margin assessment in the French nuclear industry", *Proceedings of the 17th International Conference on Structural Mechanics in Reactor Technology (SMiRT 17)*, Prague, Czech Republic, 2003.

[ARD 09] ARDILLON E., MEISTER E., FAIDY C., "Safety margins in mechanical integrity assessment for passive NPP components", *Proceedings of the 20th International Conference on Structural Mechanics in Reactor Technology* (*SMiRT 20*), Espoo, Finland, 2009.

[ARD 10] ARDILLON E., CHATEAUNEUF A., PROSKE D. *et al.*, SRA into SRA: Structural Reliability Analyses into System Risk Assessment, Paper, Det Norske Veritas, Bærum, Norway, 2010.

[ARD 13] ARDILLON E., "Time-variant structural reliability: some theoretical considerations for industrial practice", *Actes de l'ICOSSAR no. 11 (International Conference on Structural Safety And Reliability)*, New York, USA, 2013.

[ARD 15] ARDILLON E., LEMAIRE M., CHATEAUNEUF A., LANNOY A. *et al.*, "*Méthode FORM*", "Méthode" paper by 'IMdR, document produced by the IMdR Sécurité et Sureté des Structures work group, available at: https://www.imdr.eu/shop_818-47328-5300-835/fiche-methode-methode-de-fiabilite-du-premier-ordre.html, 2015.

[ARD 18] ARDILLON E., BRYLA P., DUMAS A., "A semi-probabilistic approach for optimizing quantiles in the diagnoses of hydropower penstock pipes", *Proceedings of the 54th ESReDA Seminar*, Nantes, France, 2018.

[ARD 22] ARDILLON E., YANG X., "Optimizing the maintenance of steel components in French nuclear power plants by a probabilistic approach", *Proceedings of the ICOSSAR no. 13 (International Conference on Structural Safety and Reliability)*, Shanghai, China, 2022.

[AU 01] AU S.-K., BECK J. L., "Estimation of small failure probabilities in high dimensions by subset simulation", *Probabilistic Engineering Mechanics*, vol. 16, pp. 263–277, 2001.

[BAU 17] BAUDIN M., DUTFOY A., IOOSS B. *et al.*, "OpenTURNS: an industrial software for uncertainty quantification in simulation", in GHANEM R., HIGDON D., OWHADI H. (eds), *Handbook Of Uncertainty Quantification*, Springer, Berlin, Germany, 2017.

[BAR 11] BAROTH J., SCHOEFS F., BREYSSE D. (eds), *Fiabilité des ouvrages – Sûreté, variabilité, maintenance, sécurité*, Hermès-Lavoisier, Paris, France, 2011.

[BEA 11] BEAUFILS R., MEISTER E., ARDILLON E., "Using a probabilistic approach in the brittle fracture deterministic integrity assessment of the nuclear reactor pressure vessel", *PVP2011-57528, ASME 2011 Pressure Vessels and Piping Conference*, ASME, Baltimore, USA, 2011.

[BEN 06] BEN-HAIM Y., *Info-Gap Decision Theory: Decisions under Severe Uncertainty*, Elsevier, Cambridge, MA, USA, 2006.

[BOU 11] BOURINET J.-M., DEHEEGER F., LEMAIRE M., "Assessing small failure probabilities by combined subset simulation and support vector machines", *Structural Safety*, vol. 33, no. 4, pp. 343–353, 2011.

[BOU 18] BOURINET J.-M., Reliability analysis and optimal design under uncertainty – Focus on adaptive surrogate-based approaches, HDR Thesis, Université Clermont-Auvergne, 2018.

[BRY 20] BRYLA P., ARDILLON E., DUMAS A., "Probabilistic models for penstock integrity assessment", *Proceedings of the ESREL 2020 Conference*, Venice, Italy, 2020.

[BS 15] BS 7910:2013+A1:2015, Guide to methods for assessing the acceptability of flaws in metallic structures, Guide, British Standard Institute, 2015.

[CAL 01] CALGARO J.-A., GULVANESSIAN H., "Management of reliability and risk in the Eurocodes System", *Proceedings of International Conference IABSE*, Malta, 2001.

[DAU 12] DAUTREME E., BEAUFILS R., REMY E., *et al.* "Benchmark campaign on the uncertainty treatment analysis tools developed for French nuclear reactor pressure vessels integrity assessment", PVP2012-78289, *ASME 2012 Pressure Vessels and Piping Conference*, ASME, Toronto, Canada, 2012.

[DUT 14] DUTFOY A., BAUDIN M., "OpenTURNS, logiciel opensource pour le traitement des incertitudes dans un contexte industriel", *Actes du 19ème congrès* $\lambda\mu$, Dijon, France, 2014.

[ECH 14] ECHARD B., GAYTON N., LEMAIRE M. *et al.*, "A combined importance sampling and Kriging reliability method for small probabilities with time-demanding numerical models", *Reliability Engineering and System Safety*, vol. 111, pp. 232–240, 2014.

[ELH 13] EL HAMI A., RADI B., *Incertitudes, optimisation et fiabilité des structures*, Hermès-Lavoisier, Paris, 2013.

[FAI 09] FAIDY C., HUERTA A., "PROSIR and LWR reactor vessel safety margins", Paper no. 2597, *Proceedings of the 20th International Conference on Structural Mechanics in Reactor Technology (SMiRT 20)*, Espoo, Finland, 2009.

[FRE 47] FREUDENTHAL A.M., *The safety of structures, Paper, ASCE Transactions*, vol. 112, 1947.

[GHA 91] GHANEM R.G., SPANOS P.D., *Stochastic Finite Elemnts: A Spectral Approach*, Springer, Berlin, Germany, 1991.

[HUC 18] HUCHET Q., Kriging based methods for the structural damage assessment of offshore wind turbines, PhD Thesis, Université Clermont Auvergne, Clermont-Ferrand, France, 2018.

[LAN 18] LANNOY A., LEMAIRE M., DELAGE A. (eds), *La Fiabilité en mécanique. Des méthodes aux applications*, Presses des Mines, Paris, France, 2018.

[LEL 17] LELIÈVRE N., BEAUREPAIRE P., MATTRAND C. *et al.*, "La famille des méthodes AK pour l'analyse de fiabilité : synthèse des réentes avancées", *Actes du 23ème Congrès Français de Mécanique*, Lille, France, 2017.

[LEM 09] LEMAIRE M., CHATEAUNEUF A., MITTEAU J.-C., *Structural Reliability*, ISTE Ltd, London, and John Wiley & Sons, New York, USA, 2009.

[LEM 18] LEMAIRE M., ARDILLON E., "Analyses de fiabilité des structures", *Bibliothèque Virtuelle de l'AFNOR*, 2018.

[LÉV 49] LÉVI R., "Calculs probabilistes de la sécurité des constructions", *Annales des Ponts et Chaussées*, Paris, France, 1949.

[MAD 06] MADSEN H.O., KRENK S., LIND N.C., *Methods of Structural Safety*, Dover Publications, Inc., Mineola, NY, USA, 2006.

[MAY 26] MAYER M., *Die Sicherheit der Bauwerke und ihre Berechnung nach Grenzkräften anstatt nach zulässigen Spannungen*, Springer Verlag, Berlin, Germany, 1926.

[MEL 17] MELCHERS R.E., BECK A.T., *Structural Reliability Analysis and Prediction*, 3rd ed., Wiley, Hoboken, NJ, USA, 2017.

[MET 53] METROPOLIS N., ROSENBLUTH A.W., ROSENBLUTH M.N. *et al.*, "Equations of state calculations by fast computing machines", *Journal of Chemical Physics*, vol. 21, no. 6, pp. 1087–1092, 1953.

[RAC 79] RACKWITZ R., FIESSLER B., "Structural reliability under combined random load sequences", *Computers and Structures*, vol. 9, pp. 489–494, 1979.

[RCC 00] RCC-M, Design and construction rules for mechanical components of PWR nuclear islands, Document, AFCEN, 2000.

[ROC 08] DE ROCQUIGNY E., DEVICTOR N., TARANTOLA S., *Uncertainty in Industrial Practice – A Guide to Quantitative Uncertainty Management*, Wiley, Hoboken, NJ, USA, 2008.

[RSE 05] RSE-M, In-service inspection rules for the mechanical components of PWR nuclear island, 1997 edition and addendum 2005, AFCEN, Paris, 2005.

[SOB 01] SOBOL' I.M., "Global sensitivity indices for nonlinear mathematical models and their Monte Carlo estimates", *Mathematics and Computers in Simulation*, The Second IMACS Seminar on Monte Carlo Methods, vol. 55, pp. 271–280, 2001.

[SUD 05] SUDRET B., DEFAUX G., PENDOLA M., "Time-variant finite element reliability analysis – Application to the durability of cooling towers", *Structural Safety*, 27, 93–112, 2005.

5

Probabilistic and Statistical Modeling for the Reliability of Industrial Equipment

Assessing and predicting the reliability of industrial equipment in operation on the basis of operational experience feedback data requires developing probabilistic and statistical models reflecting the nature of the equipment and the physical phenomena studied, the type of data collected and the random character of failures or degradations observed. This chapter aims to introduce the main models used in the domain of the nuclear production of electricity by emphasizing the diversity of approaches inherent to the variety of industrial needs and constraints and the complexity of the real world.

5.1. Introduction

A mathematical model aims to convey and report a phenomenon in mathematical language, that is, with the help of equations, functions, probability distributions, graphs, etc., whatever its nature (physical, biological, economic, etc.). A model is therefore a means of representing a real-world situation using mathematics. Stochastic modeling aims to represent a phenomenon's random variations, resulting from a lack of knowledge (called epistemic uncertainty) or an irreducible intrinsic natural variability of the phenomenon (hence called random uncertainty). Within stochastic modeling, we distinguish probabilistic modeling from statistical modeling. The one is conceptual and the other experimental. While the first seeks to give a formal framework and to define mathematical models of chance and uncertainty and to study their general properties, the second has the goal of comparing probabilistic models to reality, in particular to experience and observed data, in order to choose, fit and validate models, and to use them to make predictions, test hypotheses or make decisions.

Chapter written by Emmanuel REMY.

This chapter offers a short introduction to the probabilistic and statistical models most commonly used in the context of the reliability of industrial equipment in operation, in particular in the domain of the nuclear production of electricity. It has a mainly methodological aim, complementing existing international standards on statistical methods and tools for evaluating reliability [INT 97, INT 01a, INT 01b, INT 04, INT 06a, INT 07, INT 08, INT 13, and INT 16] or reference works such as monographs [KAL 11], [LAW 03] and [MEE 98]. It thus very much adopts an engineering perspective, avoiding any advanced mathematical formalization and focusing on the main principles. It is obviously not intended to be exhaustive, but sketches a fairly broad panorama of many bibliographic references for the reader who wishes to tackle some subjects in depth. By "equipment", we mean here the object needed to fulfill a function, in operation (the domain of reliability in design is not therefore addressed here [WOO 19]) making it possible to guarantee the different elementary technical functions that participate in a system's process functions: at best, we therefore place ourselves at the level of the functional group of the equipment breakdown described in Figure 2.2 of Chapter 2. Consequently, the systems studied in probabilistic safety analysis (see Chapter 3) are not related to the models presented in the following sections. Moreover, unlike the approaches developed for structural reliability (Chapter 4), the models introduced in this present chapter rely exclusively on data, whether taken from "equipment" operational experience feedback or from expertise, as described in sections 2.2, 2.4 and 2.5 of Chapter 2. It should be noted that the machine learning algorithms developed to process heterogeneous and big data, such as those taken from real-time equipment monitoring sensors, are addressed in Chapter 7 dedicated to *big data*. Finally, even though they are not specific to the domain of the nuclear production of electricity, the models displayed here certainly present the particularities inherent to this industrial sector, such as the rarity of events whose probability of occurrence we seek to evaluate, the need to use all the sources of information available to their full value or the demands of evaluating estimation and prediction uncertainty.

5.2. Some general preliminary remarks

Developing reliability models supposes that the modeler has specified at the outset the "required function", included in the literal definition of reliability given by standard [ASS 18]: "ability of an item to perform a required function under given conditions for a given time interval". It should therefore characterize precisely the unwanted event (hence the failure criterion, see section 2.5.3 of Chapter 2) whose probability of occurrence we seek to determine from the available data. If we take the example of a degradation mechanism such as corrosion cracking, several failure criteria can be defined and studied depending on a selected level of degradation that should not be exceeded, which will qualify the inability of the structure to satisfy its function: crossing the degradation threshold fixed by operation rules (which will trigger intervention to appraise the health state of the component as a preventive measure), crossing of the threshold defined as the safety criterion by the regulations (which will lead to stopping the component and a corrective action) or crossing of the final cracking threshold (which could lead to damage to the passive component's mechanical integrity).

The modeler should also consider the nature of the equipment studied, since it will, on the one hand, guide the objectives of the reliability study, and, on the other hand, condition the type of data handled and, *de facto*, the probabilistic and statistical modeling to adopt (see sections 5.4, 5.5 and 5.6). This is what the following examples seek to illustrate.

For a non-repairable active component such as a continuous flow measurement sensor, reliability can be directly assimilated to the component's duration of operation up until its failure, since it is no longer able to fulfill carry out its function once it breaks down (it is then out of service and generally replaced by another component ensuring the same mission). Assessing the reliability of this component can serve to estimate its expected lifetime in real operating conditions (possibly quite different from those that served to qualify it), to provide input data for "system" models such as those that support probabilistic safety analyses (see Chapter 3), to best set the size of the store of spare sensors or to compare different sensor technologies with one another (or any other physical factor that potentially influences reliability, whether permanent static characteristics, such as the choice of design, or dynamic scalable characteristics, such as the usage profile or environmental conditions). Operational (experience) feedback (OF) comes from the collection of operating data on the family of sensors studied. It is formed of "negative OF", made up of sensors that have broken down, but also of "positive OF" provided by sensors that have never experienced failure (which are called the survival or right-censored data – see section 2.6.2 of Chapter 2). Depending on the mesh on which the data are collected in the information system, we have either individual information sensor by sensor, with their own operating duration, or aggregated information at family level (we know for example the annual number of breakdowns experienced among all the sensors and a sensor's average annual duration of operation, with no further precision). In the first case, the modeler will consider the operating durations of the sensors such as realizations (or outcomes) of the positive continuous random variable, written T, to which must be linked a probability distribution [MEE 98, INT 16], defined in equivalent fashion by a cumulative distribution function:

$$F_T(t) = \mathbb{P}(T \leq t) \qquad [5.1]$$

(i.e. the probability that the component will break down before the operating time $t \geq 0$), a probability density function:

$$f_T(t) = \frac{dF_T(t)}{dt} \qquad [5.2]$$

(which can be roughly assimilated with the relative frequency of the time-to-failure depending on the operating duration or, in other words, with the instantaneous probability that the component breaks down exactly at the end of the operating time t), a reliability function (also called survival function):

$$S_T(t) = \mathbb{P}(T > t) = 1 - F_T(t) \qquad [5.3]$$

(i.e. the probability that the component is still in good working order at the end of the operating time t), an instantaneous operating failure rate function:

$$h_T(t) = \lim_{dt \to 0} \frac{\mathbb{P}(T \leq t + dt | T > t)}{dt} = \frac{f_T(t)}{S_T(t)} \qquad [5.4]$$

(which is interpreted as the propensity of the component that has been operating correctly for a duration t to break down at the following instant $t + dt$ – this function is also called conditional failure rate, force of mortality, hazard function, hazard rate or instantaneous failure rate) or a cumulative hazard function:

$$H_T(t) = \int_0^t h_T(u)du = -\ln(S_T(t)). \qquad [5.5]$$

In the second case of data aggregated on the component population mesh, modelers will seek to know the breakdown occurrence dynamic for the family of sensors: to do this, they will rely on the evolution of the positive random variable, although this time the variable will be discrete, counting the total number of breakdowns observed for all the sensors, which we call a stochastic counting process, written as N_t. This probabilistic object is fully defined [COO 07, INT 16] by its failure intensity function $\lambda(t)$, which conveys the propensity of observing a failure at the following instant $t + dt$, given the history of breakdowns during the operating time t of all the sensors.

In the two examples, defining a model amounts generally to giving a precise analytical form to the previous functions $F_T(.), f_T(.), S_T(.), h_T(.), H_T(.)$ or $\lambda(.)$ (we ordinarily choose a family of probabilistic models among the most standard proposed in the literature, see section 5.4), involving parameters, written as $\boldsymbol{\theta}$, which most of the time are unknown in practice and which we seek to fit as well as possible in view of the information provided by the OF data and other knowledge that may be available (see sections 5.5 and 5.6).

Now, let us consider a non-repairable active component such as a relay. Its reliability is expressed as a function of the number of handovers it can carry out before breaking down. The OF is therefore formed of the number of actuations before failure counted in the population of relays considered and the modeler will use a positive discrete random variable (integer) to express the reliability of the component. It will be written as N. Its probability distribution is given by the unitary probabilities $\mathbb{P}(N = n), \forall n$ integer and its failure rate function can be expressed by $h_N(n) = \mathbb{P}(N = n | N > n - 1)$ (see [BRA 01]).

For a non-repairable active component on standby, such as a flapper, which operates only on demand, reliability depends on its aptitude to oppose the passage of fluid when it is activated. It is therefore characterized by the flapper's probability of failure on demand [INT 10], written as $\gamma(t)$, that is, the probability that the component will not be able to execute its function at moment $t \geq 0$ when the demand on it is made. To approximate this probability of failure at demand, the modeler can rely on the OF acquired on the family of flappers analyzed, which consists of the evolution, over the operating duration of the components, of the cumulative number of failures at demand and of the total number of demands made on all the flappers. If relevant, they can also benefit from the information collected during possible periodic tests, carried out for preventive reasons on the components precisely to check their capacity to meet demand in case of need.

Let us now take the example of a repairable piece of equipment such as a feedwater pump, made of sub-components that are not repairable or difficult to repair (filter, tank, etc.) and on which corrective maintenance actions (following the occurrence of a breakdown) or preventive maintenance actions (supposed) programmed (to avoid future damage) are carried out. The reliability of this piece of equipment is the result of the combination of two components: on the one hand, its intrinsic characteristics (design, manufacture, material, etc.), and, on the other hand, the effect of maintenance tasks (repair of a failing sub-component of the piece of equipment will not have the same impact as fine-tuning or lubrication). Quantifying the reliability of this piece of equipment makes it possible to check the adequacy or relevance of the preventive maintenance program in force or to help optimize, within the safety constraints, the timetable for preventive maintenance actions (see [BAR 65]). The OF is formed of operation and maintenance data, that is, of all the equipment's operating duration (and potentially other representatives of the same family) between two consecutive maintenance operations, of the corrective or preventive type, of each of its actions, and of the nature of the latter (replacement, repair, tightening, lubrication, testing, hydraulic testing, etc.). The modeler will seek to represent the evolution dynamic (either upward or downward) of the lengths of the time intervals between two successive events. To do this, it will rely on the stochastic process describing the evolution of the continuous positive random variable corresponding to the operating duration between two maintenance interventions. Given the equipment's breakdown history and preventive maintenance actions over the operating time t, the process's intensity function, $\lambda(t)$, gives the propensity of the equipment to undergo failure and to be repaired correctively at the following instant $t + dt$ [COO 07, INT 16]. To define this intensity function, the modeler can combine an initial model representing the device's intrinsic behavior (if no preventive maintenance was carried out), of the type used for a non-repairable component, with a second model considering the impact of the different maintenance actions on the equipment's behavior.

Finally, for a passive component (structure) that undergoes a degradation mechanism, such as a water storage tank subject to corrosion cracking, reliability is linked to the period in which the degradation mechanism initiates, then to its propagation kinetic, since they both determine the duration of operating at the end of which the crack will have reached the degradation level that must not be exceeded, synonymous with failure. The OF is formed of dated information on the tank's state of degradation, collected over the course of its use (generally during inspections). Assessment of deterioration may be either qualitative, such as a visual inspection report specifying a "healthy state", "degraded state", "much degraded state" or "failed state", or quantitative, such as the depth of cracks measured by ultrasonic or eddy current testing. From these data, the modeler can, for example, envisage a model pairing the initiation and propagation phases of the degradation: the first will be represented by a positive continuous random variable, written as T_a, and the second by a stochastic process, written as D_t, that will characterize the dynamic of transition from one state of degradation to the other (for a qualitative assessment of the damage) or of the evolution of propagation (in the case of a quantitative measurement of the degradation). The main indicator of interest will then be the random operating time, written as T_ρ, at the end of which the degradation will have reached either $\mathfrak{E}_f$ the "failed state" in the qualitative case: $T_\rho = \inf_{t \geq T_a}\{t | D_t \in \mathfrak{E}_f\}$, or the allowable degradation limit threshold ρ in the quantitative case: $T_\rho = \inf_{t \geq T_a}\{t | D_t \geq \rho\}$.

To conclude these general remarks, it seems important to present briefly the overall methodological approach to which a statistical reliability study belongs. This approach relies on four stages, all as important as one another.

1) The first step consists of analyzing the industrial problem and formalizing it in terms of reliability to define, on the one hand, the question asked (what are the unwanted event and the associated failure criterion), and, on the other hand, the quantity of interest to evaluate to have tangible elements to help decision-making (average lifetime of a non-repairable component, number of breakdowns expected during the following operating cycles of a family of components, efficiency of a preventive maintenance program, influence of an operating or environmental

parameter on the evolution of a degradation, etc.). During this stage, it makes sense to reflect on "ideal data", which is theoretically needed to properly answer the question asked, then to look at the "real data" which we hope to be able to have access in concrete form. In case of significant deviation between theoretical forecast and observable practice, it is then necessary to study the relevance and effort needed to collect additional knowledge with regard to the level of the decision to be taken. At this stage, it makes sense to emphasize again the fact that, generally, the data to be collected are not limited to failure events and their dates (or associated operating durations) alone. Depending on the objectives of the study, and so as to guarantee correct interpretation and good use of data, the collection of information should also focus (among other things) on the nature of the equipment studied, the failure modes (necessary for probabilistic safety analyses, see Chapter 3), the degradation mechanisms, data on proper functioning (the positive OF provided by survival cases), maintenance operations (corrective or preventive, content of tasks, costs incurred, etc.), the usage profile (annual average number of operating hours or demands, shutdowns, etc.), potentially influential physical factors, etc. We should also note the benefit of free-form summaries, written, for example, following maintenance interventions carried out on components, which provide, through written text, useful information to help the understanding and the relevance of the data and its validation.

2) Once the data are collected and validated (especially with regard to their accuracy and their relevance for the problem posed, see section 2.5.4 of Chapter 2) and following an exploratory analysis of these data, the second stage aims to build one (or more) probabilistic model(s) describing the unwanted event studied (see section 5.4), taking account of the question posed, the nature of the information collected and the number of parameters to consider (if we wish, for example, to involve several other factors in the probabilistic model, in particular physical factors influencing the occurrence of the undesirable event). This step is the translation of the industrial problem into probabilistic terms.

3) The third stage, linked to the previous one, consists of estimating the parameters $\boldsymbol{\theta}$ of the model(s) developed at phase 2) (see sections 5.5 and 5.6), then of evaluating numerically the quantities of interest sought, which derive directly from the model and from the estimated value of $\boldsymbol{\theta}$, written as $\widehat{\boldsymbol{\theta}}$. While step 2) is of probabilistic nature, stage 3) is of statistical nature. It is in the course of this stage that a number of constraints inherent to the industrial context will emerge: data from OF, low in number, highly censored, or indeed lacking, will, for example, require the use of particular statistical models (Bayesian approach – see section 5.6, expectation-maximization algorithm EM [BAC 98], etc.). Similarly, assessment of the statistical estimation uncertainty relating to the quantities of interest will perhaps require implementing re-sampling techniques (such as the bootstrap [EFR 94] – see section 5.5), since the results of asymptotic convergence of statistical estimators

of $\boldsymbol{\theta}$ are not necessarily applicable with sizes of databases encountered. Both stages 2) and 3) are highly dependent on one another and condition the applicability of the model developed. In fact, this latter should be adapted to the objectives of the study and to the data available and should be possible to estimate robustly to provide quantities of interest true to reality for fully informed decision-making.

4) This last step aims to validate the model established previously with the aid of criteria, statistical tests or field expert considerations, so as to check that it is consistent with the knowledge that has served to build it, but also to compare the model with other models that may be in competition (see section 5.7). The goal here is to find a robust, parsimonious model (i.e. with the fewest parameters possible), with a strong predictive capability, providing the best representation possible of the unwanted event and the associated uncertainties.

5.3. Non-parametric approaches

Before seeking to establish a model, it is vital to explore the data available. This preliminary work targets many objectives: to familiarize ourselves with the data (the size of the database, the variables studied, their value scales, the number of components and failures, the temporal horizon covered by the data, etc.), to detect potential sources of difficulty (missing, erroneous or atypical values, under-representation of a category, etc.), to summarize the data using some synthetic numerical indicators to better understand it (frequency tables, means, standard deviations, quantiles, correlations, censoring rates, etc.), visualizing data thanks to graphical representations to identify trends or typologies in observations (histograms, box plots, scatter plots, etc.), or to help make diagnosis and formulate hypotheses (symmetrical distribution, link between different factors, similar behavior between two families of equipment, constant failure rate function, etc.) that could be verified with the help of models or statistical tests (see section 5.7).

In this context, non-parametric approaches are very useful tools. Unlike parametric statistics (see sections 5.4–5.6), non-parametric techniques do not rely on models described by a number of finite parameters $\boldsymbol{\theta}$ (such as those presented in section 5.4). Consequently, they only use strictly relevant information provided by the data and make no hypothesis on the form, nature or type of probability distribution or stochastic process that generated the observations: they therefore authorize all possibilities, which is especially advantageous, above all in the framework of a preliminary exploratory study.

The most well-known example of a non-parametric approach in reliability is certainly the estimator of the survival function $S_T(t)$ proposed by Kaplan and Meier

[KAP 58]. This estimator (also called product limit estimator) measures, without postulating a model and in the case of an OF collected individually for each non-repairable component in a family, the fraction of components still in good working order after the operating time $t \geq 0$, making it possible to take account of potential right-censored data (i.e. the components that have not failed). This point estimation of the reliability function (which maximizes the likelihood function, see section 5.5) can be accompanied by a confidence interval, the width of which helps to quantify the precision of the estimation in view of the information provided by the data (see section 5.5). This way of approaching the function $S_T(t)$ gives an initial glimpse of the evolution of the reliability of the components during operation and makes it possible easily, for example, to compare several families with one another, as Figure 5.1 illustrates, and so to highlight the potential need to consider influencing factors in the modeling (see section 5.4.3).

Many other non-parametric tools dedicated to the domain of equipment reliability exist, among them the reliability function estimator using life table analysis for data aggregated at the level of a family of non-repairable components [LAW 03], the non-parametric test of the log-rank to compare the survival functions of at least two families of components [BAG 11b], the Nelson-Aalen or Breslow estimators of the cumulative failure rate function H(t) [KAL 11] or the kernel estimators of the instantaneous failure rate function $h_T(t)$ of the lifetime T of non-repairable components or the intensity function $\lambda(t)$ of some stochastic processes for repairable equipment [GAM 11]. Bracquemont [BRA 01] presents several non-parametric approaches in the case where discrete data are manipulated, such as those taken from the observation of circuit breakers or relays (see sections 5.2 and 5.4.2).

Despite their undeniable benefit (no *prior* knowledge on the data generation process, faithful representation of data, "more general" models so ones presenting some "robustness", smooth visual aspect of the estimated functions, etc.), non-parametric approaches present two main limits, above all in an industrial application context: on the one hand, for an equivalent estimation precision, they require more data than parametric approaches; on the other hand, they do not lend themselves very well to the exercise of extrapolating beyond the data observed (in our case, beyond the operating times of the equipment forming the OF). This is why they are rarely used alone and benefit from joint use with parametric models, that make it possible to remove some obstacles.

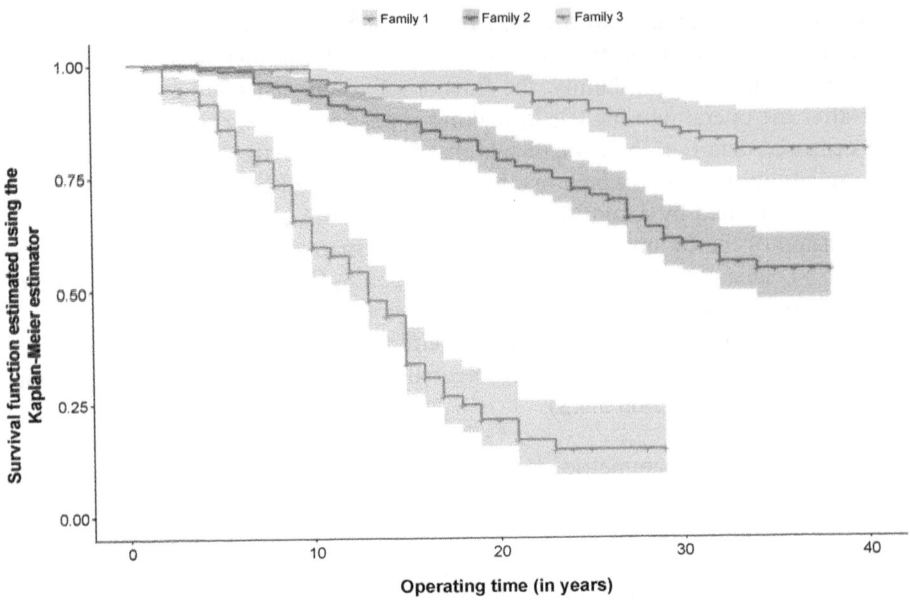

Figure 5.1. *Examples of Kaplan–Meier estimations of the survival functions of three families of non-repairable components (nuclear heat exchangers, which have the function of making it possible to collect fluid samples). The envelopes correspond to 90% bilateral asymptotic confidence intervals. Since no curve overlaps, we can reasonably infer that the three families of components have different reliability functions. This graph was generated using the "survival" package from R software environment. For a color version of this figure, see www.iste.co.uk/lannoy/nuclear.zip*

5.4. Parametric models

5.4.1. *Introduction*

Defining a parametric model comes down to supposing that the random phenomenon that has generated the data can be represented correctly by a family of probability distributions or stochastic processes, indexed by a finite set of parameters. A whole variety of families of parametric models exists in literature, as we will see in this section, and it is for the modeler to adopt the one best adapted to the nature of the equipment studied, to the type of data handled and to the objectives targeted in the study (see section 5.2), ideally oriented by the conclusions of the exploratory analysis of the data (see previous section). Once a family has been chosen, frequentist or Bayesian statistical approaches (see sections 5.5 and 5.6) are used to "complete" the model, specifying the values or distributions of its parameters in view of all the information available. Several families of distributions or processes could evidently be suitable at the outset: modelers must therefore test all the candidate models, then select only one on the basis of statistical criteria or professional engineering considerations (see section 5.7). They may also decide to build their own parametric model, specifically developed for their problem, if none of the most standard fully meet expectations. Whatever the approach chosen, a model should, if possible, display the following, sometimes opposing, qualities. It should:

– be parsimonious (or sparse), that is, with a good compromise between, on the one hand, a large enough number of parameters for the model to be flexible, for it to fit the data properly[1], and, on the other hand, a sufficiently reduced number of parameters for the model to be relatively insensitive to the details of the data and for it to display good generalization[2];

– be robust, that is, relatively insensitive to measurement errors or to slight deviations from working hypotheses;

– be easy to manipulate (easy numerical calculations and stability of results);

– have good predictive power (the model should forecast new data with good precision);

– and be easily interpretable.

This modeling work requires a close collaboration between the mathematical modeler and the field expert or manager, who will often be the users of the results from the reliability study.

5.4.2. *Some models adapted to non-repairable components*

For a non-repairable active component that operates continuously and for which in the OF there is an individual information set on the mesh for each component of the population studied, several families of probability distributions exist in literature for modeling the positive continuous random variable T, characterizing the component's lifetime. The most well known is certainly the Weibull distribution, presented in detail in [RIN 08] and classically parametrized by $\boldsymbol{\theta} = (\eta, \beta)$, with $\eta > 0$ the scale parameter (in the same unit of time used to express T in the OF, for example years) and $\beta > 0$ the shape parameter (without a unit). It will be written $\mathcal{W}(\eta, \beta)$ in the following. This distribution relies on the following expression of the cumulative distribution function:

1 We therefore say that a model displays a "weak bias", that is, a "slight deviation from the data".

2 We therefore say that a model displays a "low variance": it is "resilient".

$$F_{T,\mathcal{W}}(t|\boldsymbol{\theta}) = \mathbb{P}(T \leq t) = 1 - \exp\left(-\left(\frac{t}{\eta}\right)^{\beta}\right), \forall t \geq 0. \quad [5.6]$$

It displays the following properties, which explain its popularity.

– Its distribution can take very varied forms depending on the values of the two parameters (η, β), thus allowing good flexibility in terms of modeling (see Figure 5.2).

– It is connected to many physical or biological models [RIN 08], such as the weakest link model or the broken stick model used to describe the abundance of different animal species in a habitat, so conferring upon it a particular grounding in reality.

– With Gumbel and Fréchet, it is one of the three families of limit laws in extreme value theory, which provides it theoretical foundation.

– Its instantaneous failure rate function takes a simple and easily interpretable expression:

$$h_{T,\mathcal{W}}(t|\boldsymbol{\theta}) = \lim_{dt \to 0} \frac{\mathbb{P}(T \leq t+dt|T>t)}{dt} = \frac{\beta}{\eta}\left(\frac{t}{\eta}\right)^{\beta-1}, \forall t \geq 0. \quad [5.7]$$

Thus, depending on the value of the shape parameter and as Figure 5.2 illustrates, the failure rate function of the Weibull distribution decreases according to the component's operating time when $0 < \beta < 1$ (which characterizes what is called a rejuvenation, a debugging, an infant mortality or early failures: the longer the component has functioned correctly, the lower the probability that it will break down over the course of the few instants that follow its operating duration t), or constant over time when $\beta = 1$ (we then stumble across the exponential distribution with rate parameter $\lambda = \frac{1}{\eta} > 0$, written $\mathcal{E}(\lambda)$, which presents the particularity of being "memoryless": the past has no bearing on the future behavior of the component – the probability of breakdown over the course of the few instants that follow the operating time of the component t is the same whatever the operating duration it has attained – we say that the component is in its useful life), or increasing depending on the operating duration when $\beta > 1$ (we therefore speak of ageing or wear out: the longer the component has functioned correctly, the higher the probability it will break down over the course of the few instants that follow its operating time t). The Weibull model therefore makes it possible to model one of the three periods of the famous bathtub curve that classically characterizes the lifecycle of a component. Under the hypothesis of ageing, we generally distinguish the three following cases: when $1 < \beta < 2$, the failure rate function is increasing and concave; when $\beta = 2$, it is linear; and when $2 < \beta$, it is increasing and convex, i.e. the higher the shape parameter, the greater the intensity of the statistical ageing (we often speak in this case of "accelerated ageing").

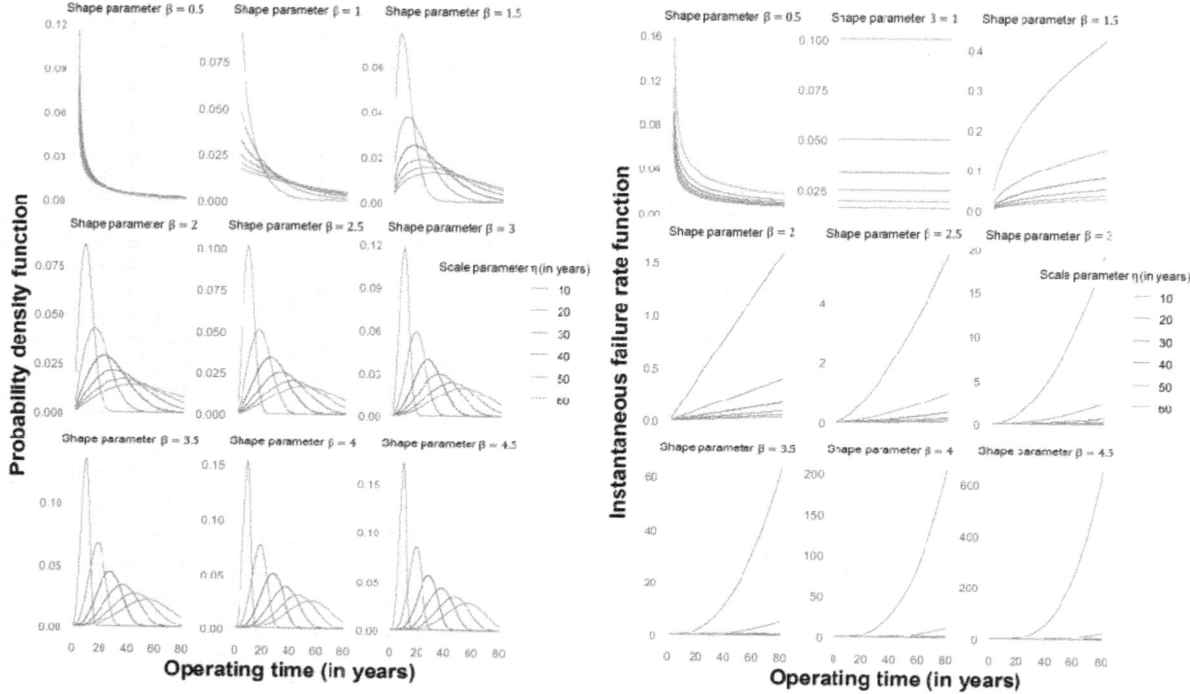

Figure 5.2. *On the left: probability density functions of Weibull distributions for values of parameters* $\eta = \{10; 20; 30; 40; 50; 60\}$ *(in years) and* $\beta = \{0.5; 1; 1.5; 2; 2.5; 3; 3.5; 4; 4.5\}$*; on the right: instantaneous failure rate functions of Weibull distribution for the same values of the two parameters* η *and* β *– this illustrates the flexibility of the Weibull distribution, which presents very varied behaviors depending on the values of its parameters; graph generated using the "flexsurv" package from R software environment. For a color version of this figure, see www.iste.co.uk/lannoy/nuclear.zip*

Other families of distributions are also used currently to represent non-repairable components which break down during operation, among these are log-normal and log-logistic distributions, gamma distribution (widely used in Bayesian statistics, see section 5.6), Pareto distribution (common in extreme value theory), inverse Gaussian distribution, Birnbaum-Saunders distribution or Gompertz distribution, which we will not detail here (see [MEE 98]).

The Weibull model can be extended to consider the various characteristics the component may present. We can cite:

– the addition of a third location parameter (in the same unit as the scale parameter) to consider a non-random respite (or initiation) period, during which a breakdown is not possible;

– linking Weibull distribution to another random variable characterizing the respite period, this considered to be random;

– the three- or four-parameter Bertholon model [BER 01], which places two failure modes that the component may undergo in competition (the first modeled by an exponential distribution, the second by a Weibull model), which makes it possible to model an instantaneous failure rate function that is non-zero at the commissioning time, indeed to cover two or three periods of the bathtub curve, for example useful life, then ageing.

The Bertholon model is a good transition for broadening this presentation to the modeling framework known as competing risk models, adapted to the case where the component studied is subject to a number (assumed to be known, we will denote it k) of undesirable events, such as several failure modes, degradation mechanisms or causes of breakdown. The component's reliability is not therefore characterized by a single random variable, but by k, written $T_1, \ldots, T_k$, each of them representing the operating duration at the end of which the associated undesirable event occurs. If a single one of k events is enough to lead to the breakdown of the component, the k events are in competition with one another and the occurrence of one event will prevent the others from happening (they will not therefore be observed: they are censored because the first event has occurred). If T represents the component's lifetime, we have $T = \min(T_1, \ldots, T_k)$: the probability distribution of T depends therefore on the distributions of the operating durations at the end of which the undesirable k events occur, and these may (for physical reasons for example) not be mutually dependent of one another. Thus, it is not enough to study the undesirable k events separately (marginally) to determine the distribution of T and it is really the joint (multidimensional) analysis of variables $T_1, \ldots, T_k$ that makes it possible to characterize fully the reliability of the component. The interested reader may refer [CRO 12] for a broader glimpse of the tools adapted to this specific case.

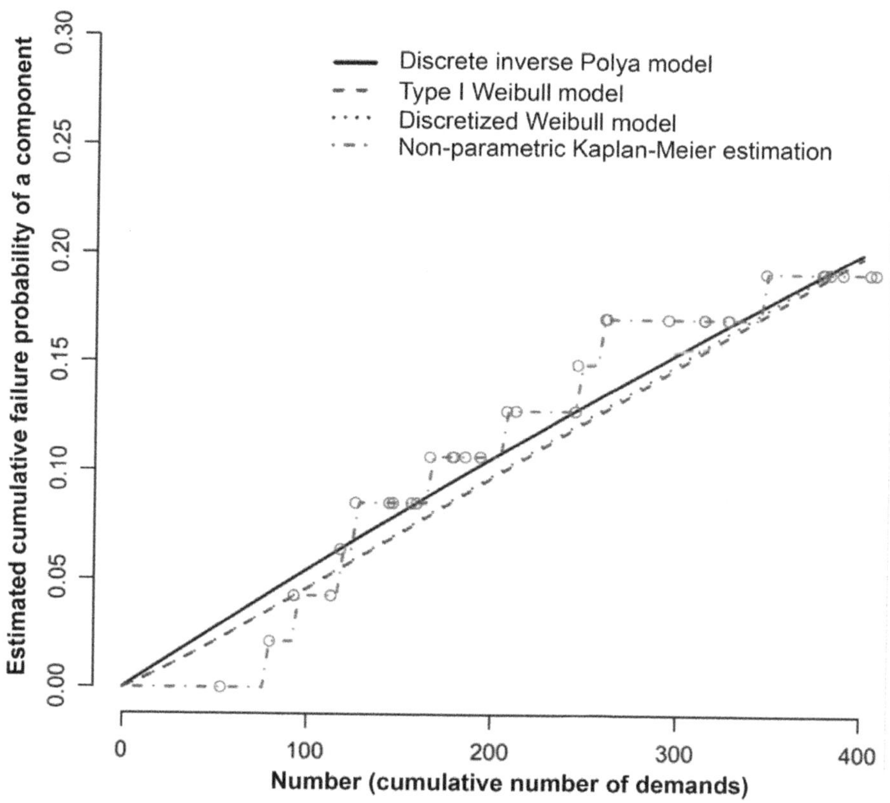

Figure 5.3. *Illustration of the cumulative distribution functions of three discrete models estimated using OF data from a family of nuclear power plant components. Note a good overlap of these curves with the non-parametric Kaplan–Meier estimation (see section 5.3), showing the good capacity of the models to fit real data – extract from [PAS 15]. For a color version of this figure, see www.iste.co.uk/lannoy/nuclear.zip*

For active non-repairable components whose life is expressed by a discrete positive random variable, such as the number N of handovers at the end of which a relay experiences a failure, the literature is more restricted. Two main families of distributions exist [BRA 01]. The first is composed of discrete distributions derived from standard continuous distributions in reliability. This transition from a continuous time to a discrete time can be made, either by constructing a similarity of expression between continuous time and discrete time for a characteristic quantity, or by discretizing the time by taking the integer part of the continuous random variable T. In this first family, we find the geometric distribution (which presents the characteristic of being memoryless, like time-continuous exponential distribution), the shifted negative binomial distribution (equivalent to the gamma distribution), types I, II and III derivations of the Weibull distribution (based respectively on the same expressions of the reliability function, the failure rate function and the scale parameter of the continuous Weibull distribution), and all the classical positive continuous distributions discretized by taking their integer part. The second family of discrete distributions relies on Polya's urn scheme [EGG 23]. It consists of assimilating each of the demands on the component to a lot in an urn containing a certain number of red balls (associated with the breakdown event) and green balls (synonymous with proper functioning), with strategies to replace the balls after each draw to be defined and which will make it possible to model an evolution of the probability of breakdown at the next sampling (replacing each of the green balls sampled to the urn – we come across the classic binomial distribution, or replacing each green ball sampled and adding a number of new red balls conveying statistical ageing, etc.). For more clarifications on these models, the interested reader may refer to [BRA 01] and [PAS 15]. In particular, this last bibliographical reference (from which Figure 5.3 is taken) presents and compares the application (for the purposes of optimizing the preventive maintenance program) of Polya's, Weibull's type I and Weibull's discretized models on two families of nuclear power plant components, illustrating the applicability and industrial interest of these models.

5.4.3. *Taking account of influencing factors*

A significant problem in industrial reliability (and an increasingly promising one in a big data context – see Chapter 7) is to identify then explain the heterogeneities of behavior between several families of components or, in other words, to determine from the data whether observed factors (and available in data from OF), relating for example to a component's design, manufacture, operation or environmental conditions, can influence its lifetime.

From this perspective, two approaches can be envisaged.

– If modelers know the influencing factors, they can study each of the families of components separately, having defined these families at the outset on the basis of crossing the factors. For example, if we know that the geographical location of nuclear power plants (by the coast, on a river) impacts the lifetime of a type of component, it will be possible to carry out a distinct statistical analysis for each of the two families of components, defined by the two possible categories of reactor location. In the case of a quantitative factor, such as the carbon content of the chemical composition of the material that constitutes the equipment, we could carry out as many statistical studies as there are relevant ranges of values in the factor (e.g. "low content", "average content" and "high content"). Aside from the fact that influencing factors are not generally (all or well) known in practice, this approach is only partly satisfactory, since, on the one hand, it is not realistic to believe there will be enough observations in each family, and, on the other hand, it does not allow easy comparison between families, since none of them acts as a reference by which all the others could be calibrated.

– The other approach is to establish a model in which the distribution of the random variable T (or of a function of T) will be expressed according to the parameters that characterize the effect of the values or categories taken by each of the factors on the probability distribution. In statistics, these types of model expressing the distribution of T conditionally on $p \geq 1$ (potentially) influential factors, which will be written $\mathbf{Z} = (Z_1, \ldots, Z_p)$, are called regression models with covariates or explanatory variables. The choice of factors is in the hands of the modeler, who can rely, either at the very start of the modeling process, on field expertise or the results of a descriptive statistical study that will have allowed him/her to highlight the covariates to retain, or after constructing the model, on using statistical hypothesis tests that will allow him/her to check the influencing character of the variables considered (see section 5.7).

The Cox model [COX 72] is one of the most popular regression models with covariates. It makes it possible to model the effect of covariates acting in multiplicative fashion on the component's lifetime distribution. It consists of expressing the instantaneous failure rate function of T over time as the product of the first term $h_0(t)$, a function of the operating time $t \geq 0$, characterizing the component's baseline hazard rate function (when all the covariates are zero), and of a second term that represents the relative effect of the explanatory variables compared to the reference situation, by weighting the baseline hazard with a positive parametric link function (in this case log-linear) $\Psi(., \boldsymbol{\beta}) > 0$ of the values taken by the covariates $\mathbf{Z}$:

$$h_{T,Cox}(t, \mathbf{Z}|\boldsymbol{\beta}) = h_0(t)\Psi(\mathbf{Z}, \boldsymbol{\beta}) = h_0(t)\exp(\boldsymbol{\beta}^T\mathbf{Z}) = h_0(t)\exp\left(\sum_{k=1}^{p} \beta_k Z_k\right), \forall t \geq 0 \quad [5.8]$$

with $\boldsymbol{\beta}^T = (\beta_1, \ldots, \beta_p)$ a parameter vector (the parameters are generally unknown and should be estimated from the data – see section 5.5). In equivalent fashion, we can express the previous equation according to the survival function:

$$S_{T,Cox}(t, \mathbf{Z}|\boldsymbol{\beta}) = \left(S_0(t)\right)^{\Psi(\mathbf{Z},\boldsymbol{\beta})}, \forall t \geq 0 \quad [5.9]$$

where $S_0(t) = \exp\left(-\int_0^t h_0(u)du\right)$ is the reliability function associated with the baseline hazard rate function. The model was originally formulated in a semi-parametric framework, in the sense that Cox did not consider it useful to give a particular shape to or to estimate the baseline hazard function $h_0(.)$, which is the same for all the components at a given instant: it only seemed important to him to evaluate the failure rate ratio for two components exposed to different **Z** covariates. From this formulation and supposing that **Z** and **β** are not time-dependent, comes the proportional hazard assumption of the Cox model: for two components that differ only by one unit in only one of the covariates, we say $Z_k, 1 \leq k \leq p$, which will be supposed to be numerical, then the ratio of their failure rates is expressed by: $\frac{h_0(t)\exp(\beta_1 Z_1+\cdots+\beta_k(Z_k+1)+\cdots+\beta_p Z_p)}{h_0(t)\exp(\beta_1 Z_1+\cdots+\beta_k Z_k+\cdots+\beta_p Z_p)} = \exp(\beta_k), \forall t \geq 0$. The ratio is therefore independent of the baseline hazard rate function and the operating time and, whatever t, one of the components has an instantaneous failure rate that is $\exp(\beta_k)$ times that of the other component. If $\exp(\beta_k) > 1$, the Z_k covariate induces a multiplicative effect strictly superior to 1 on the failure rate: the component therefore presents greater statistical ageing than in reference conditions (when $Z_k = 0$). This modeling hypothesis using proportional hazards is evidently strong and it should be checked on the data (see section 5.7) to ensure that the Cox model can faithfully characterize the behavior of the components studied: to this end, there exist graphical approaches and statistical tests, for instance the Schoenfeld residuals test (see [THE 00]).

There are variants of the standard Cox model. We can mention the most widely used, which consist of giving a precise parametric shape to the baseline hazard function $h_0(t)$ or changing the link function $\Psi(\mathbf{Z}, \boldsymbol{\beta})$ (which should be positive and verify the condition $\Psi(\mathbf{0}, \boldsymbol{\beta}) = 1$).

The other classical model is called the scale-accelerated failure time (AFT) model [MEE 98, BAG 01]. It is constructed to explain the effect of a stressor on one of the families of components and relies on the following relation of the survival function:

$$S_{T,AFT}(t, \mathbf{Z}|\boldsymbol{\beta}) = S_0(t \times \Psi(\mathbf{Z}, \boldsymbol{\beta})), \forall t \geq 0. \quad [5.10]$$

Everything therefore happens as if, given the **Z** covariates, the lifetime T of the component was equal to the nominal life T_0 (without a stressor, with a reliability function $S_0(.)$) divided by $\Psi(\mathbf{Z}, \boldsymbol{\beta})$: $T = \frac{T_0}{\Psi(\mathbf{Z},\boldsymbol{\beta})}$. When the acceleration factor $\Psi(\mathbf{Z}, \boldsymbol{\beta}) > 1$, the component ages more quickly than in the reference operating conditions without stress (when all the covariates are zero). Section 5.8 is dedicated to presenting a concrete use of this family of models for a type of nuclear power plant components, showing the utility of these models.

Other models are available in the literature [THE 00], among them the stratified Cox model (when at least one of the covariates depends on the time and/or has no multiplicative effect on the failure rate and we do not wish to estimate its effect), the models with time-dependent factors or competing risk models (when the component is subject to several failure modes, see section 5.4.2 and [CRO 12]).

We will conclude this section by mentioning the example where unfortunately, the heterogeneity of behavior between the components cannot be explained by the only factors available in the OF data: there are therefore other hidden sources (or ones that cannot be reached/measured, such as manufacturing conditions) and in this case, other families of covariate models should be used, such as frailty models [WIE 10]. The most standard consists of introducing a random multiplicative term ε into the definition of the failure rate to consider the unobserved (or hidden) heterogeneity:

$$h_{T,frag.}(t, \mathbf{Z}|\varepsilon, \boldsymbol{\beta}) = \varepsilon h_0(t)\Psi(\mathbf{Z}, \boldsymbol{\beta}), \forall t \geq 0. \quad [5.11]$$

ε is the unobservable frailty term (also called the random effect), which will vary from one component to another. It will classically be supposed to be distributed according to a particular given distribution (gamma for example), the dispersion of which will quantify the heterogeneity between the components or families of components. More generally, this problem of the unobserved factor which we know affects the random variable of interest studied (in this instance the lifetime of component) makes reference in statistical literature to models called latent variable models. The interested reader will find a glimpse, in Chapter 11 of [RAM 19], of models in this category that have displayed their benefit in the field of component reliability.

5.4.4. *Imperfect maintenance models for repairable equipment*

According to standard [ASS 18], the maintenance is the "combination of all technical, administrative and managerial actions during the life cycle of an item intended to retain it in, or restore it to, a state in which it can perform the required function". Maintenance, therefore, contributes in an essential way to the operational reliability of equipment and it is a major challenge to model and evaluate quantitatively the effect of maintenance actions, so as to optimize those actions in compliance with safety, availability and cost constraints. Although corrective maintenance is carried out after fault recognition and intended to restore the piece of equipment into a state in which it can again perform the required function, preventive maintenance is carried out in accordance with established intervals of time or according to prescribed criteria based on physical conditions intended to assess and/or to mitigate degradation and reduce the probability of failure of the piece of equipment. In what follows, we will focus solely on the effect of different types of maintenance (we will neglect in particular the logistic delay and the time to carry out these actions).

We will focus, in the beginning, on the case of a repairable piece of equipment on which only corrective maintenance after failure is carried out. We will write T_1 the random variable that characterizes its intrinsic life duration in the absence of maintenance: T_1 is the operating time of the new, unmaintained device, which can be characterized by its instantaneous failure rate function $h_{T_1}(t) = \lim_{dt \to 0} \frac{\mathbb{P}(T_1 \le t+dt | T_1 > t)}{dt}, \forall t \ge 0$. We write $T_i, i \ge 1$, the repairable device's successive random operating times to failure (we also speak of recurrent events). With the hypothesis that the maintenance times are not considered, the T_i also represent instants of corrective maintenance actions. We also write N_t, $t \ge 0$, the cumulative number of failures observed on the device up until the operating time t. From a mathematical point of view, the $\{T_i\}_{i \ge 1}$, or the equivalent $\{N_t\}_{t \ge 0}$, constitute a random point process [COO 07]. It can be characterized completely by its (failure) intensity function:

$$\lambda(t) = \lim_{dt \to 0} \frac{1}{dt} \mathbb{P}\left(T_{N_{t^-}+1} \le t + dt \middle| N_{t^-}, T_1, \dots, T_{N_{t^-}}\right), \forall t \ge 0 \qquad [5.12]$$

where N_{t^-} is the cumulative number of failures strictly observed before time t. Until the first instant of failure T_1, the intensity function $\lambda(t)$ is equal to the failure rate function $h_{T_1}(t)$, which will therefore be called the initial intensity function and which will simply be written $h(t)$ below (without reference to variable T_1).

The first model, called minimal maintenance efficiency model, supposes that the effect of corrective repairs is to leave the piece of equipment in the state as it was before maintenance. With this modeling hypothesis, maintenance has the effect neither of improving, nor of deteriorating the state of the piece of equipment. This hypothesis is known as "As Bad As Old" (ABAO). The random process of undergoing failures is therefore a non-homogeneous Poisson process and its intensity function is deterministic and always equal to the initial intensity function: $\lambda_{ABAO}(t) = h(t), \forall t \ge 0$. With an initial intensity function of Weibull type (i.e. if the operating life of the new, unmaintained device T_1 follows a Weibull distribution, see section 5.4.2), the failure process is a power law process, the intensity function of which is expressed by $h(t) = h(t|\eta, \beta) = h_{T_1,\mathcal{W}}(t|\eta, \beta) = \frac{\beta}{\eta}\left(\frac{t}{\eta}\right)^{\beta-1}, \eta > 0, \beta > 0$, η and β being the scale and shape parameters of the Weibull distribution associated with T_1. If $\beta = 1$, we obtain the particular case of the standard homogeneous Poisson process [COO 07].

Conversely, the perfect maintenance efficiency model supposes that each corrective replacement leaves the piece of equipment as good as new. This hypothesis is called "As Good As New" (AGAN). At any time, the piece of equipment is equivalent to a new one that has not undergone failure (or maintenance) and put into service at the date of the last corrective maintenance $T_{N_{t^-}}$. The operating durations between the failures are then independent and have the same distribution. The failure process is called a renewal process, the intensity function of which verifies: $\lambda_{AGAN}(t) = h(t - T_{N_{t^-}}), \forall t \geq 0$. As Figure 5.4 illustrates, the failure intensity function returns to 0 (new device) after each corrective maintenance action, following the same curve as the initial intensity function.

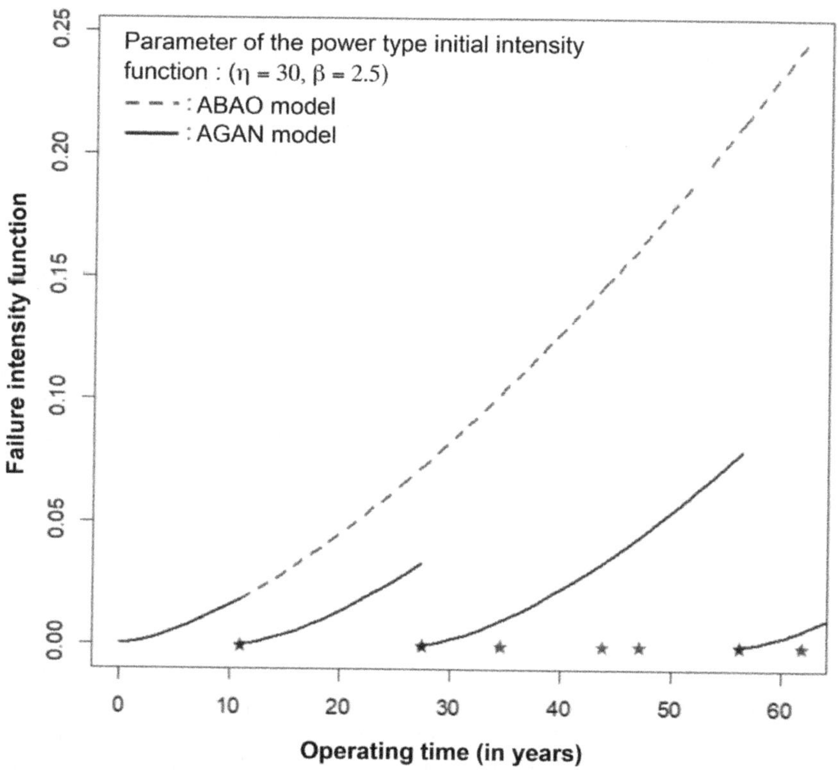

Figure 5.4. *Failure intensity functions for the two models, ABAO and AGAN, with the same power type initial intensity function with parameters ($\eta = 30\ years\,; \beta = 2.5$). The stars on the x-axis correspond to times to failure (or in equivalent fashion to corrective maintenance times under our hypothesis) simulated with both models. The failure intensity function is unchanged after an ABAO corrective maintenance, whereas it restarts at 0 after each AGAN corrective maintenance. The graph was generated with the "VAM" package from the R software environment. For a color version of this figure, see www.iste.co.uk/lannoy/nuclear.zip*

By its nature, maintenance has an effect on failure intensity. However, this effect is not necessarily to return the device to its original state: the reality lies somewhere in between these two extreme cases presented above. In the literature, models making it possible to consider an intermediary maintenance efficiency between ABAO and AGAN are known by the name of imperfect maintenance models. Many models have been suggested, the most used being Virtual Age Models (VAM) [KIJ 89]. They are characterized by a set of positive random variables $\{A_i\}_{i\geq 1}$, with $A_0 = 0$, and rely on the hypothesis according to which, after the ith maintenance, $i \geq 1$, the device behaves as new one having survived without failing until time A_i. Thus, the intensity function of a virtual age model verifies: $\lambda_{VAM}(t) = h(t - T_{N_{t^-}} + A_{N_{t^-}}), \forall t \geq 0$. The quantity $t - T_{N_{t^-}} + A_{N_{t^-}}$ is called the "virtual age of the equipment". As a particular case of the virtual age model, we find the two basic models: if $A_i = T_i, \forall i \geq 1$, the maintenance is minimal (ABAO) and if $A_i = 0, \forall i \geq 1$, the maintenance is perfect (AGAN). There are different virtual age models [DOY 04], of which the Brown–Proschan model (which supposes that the maintenance is perfect AGAN with a probability $p, 0 \leq p \leq 1$, and minimal ABAO with a probability $1 - p$) or arithmetic reduction of age (ARA) models. Among these, we find the ARA model with infinite memory, written ARA∞, which assumes that maintenance reduces the virtual age of the piece of equipment by an amount which is proportional to its virtual age just before repair. As the virtual age just before the ith failure, $i \geq 1$, was equal to $T_i - T_{i-1} + A_{i-1}$, the virtual age just after the ith maintenance is: $A_i = (T_i - T_{i-1} + A_{i-1}) - \rho(T_i - T_{i-1} + A_{i-1}) = (1 - \rho)(T_i - T_{i-1} + A_{i-1})$ where $\rho \leq 1$ is a proportionality factor (generally unknown; we will seek to determine it from OF data or initial knowledge, see sections 5.5 and 5.6). The ARA∞ model's intensity function verifies:

$$\lambda_{ARA\infty}(t) = h\left(t - \rho \sum_{j=0}^{N_{t^-}-1} (1 - \rho)^j T_{N_{t^-}-j}\right), \forall t \geq 0. \qquad [5.13]$$

The ρ parameter characterizes the efficiency of maintenance: if $\rho = 1$, the maintenance is perfect (AGAN); if $0 < \rho < 1$, the maintenance is efficient, which is generally the effect we expect to observe in practice; if $\rho = 0$, the maintenance is minimal (ABAO); and if $\rho < 0$, the maintenance is detrimental, i.e. it deteriorates the piece of equipment. These Brown–Proschan and ARA∞ models present the advantage of characterizing explicitly, on the one hand the intrinsic ageing through the initial intensity function $h(.)$, and on the other hand the effect of maintenance on the equipment's reliability via the parameter p (Brown–Proschan model) or ρ (ARA∞ model). This facilitates the interpretation of these models and their appropriation by non-specialists. As Figure 5.5 shows, we observe that, for a single initial intensity function (with intrinsic ageing, since the shape parameter β of the power type model is strictly greater than 1 – see section 5.4.2), the maintenance effect model ARA∞ with parameter $\rho = 0.7$ leads to an (operational) failure intensity function generally less than that of the same ARA∞ model, but with a lower efficiency parameter $\rho = 0.3$. It makes sense to emphasize here the fact that, in this precise case, the model presents three generally unknown parameters $\boldsymbol{\theta} = (\eta, \beta, \rho)$ that we will seek to determine with the data available from operation and from maintenance feedback or from initial knowledge (see sections 5.5 and 5.6).

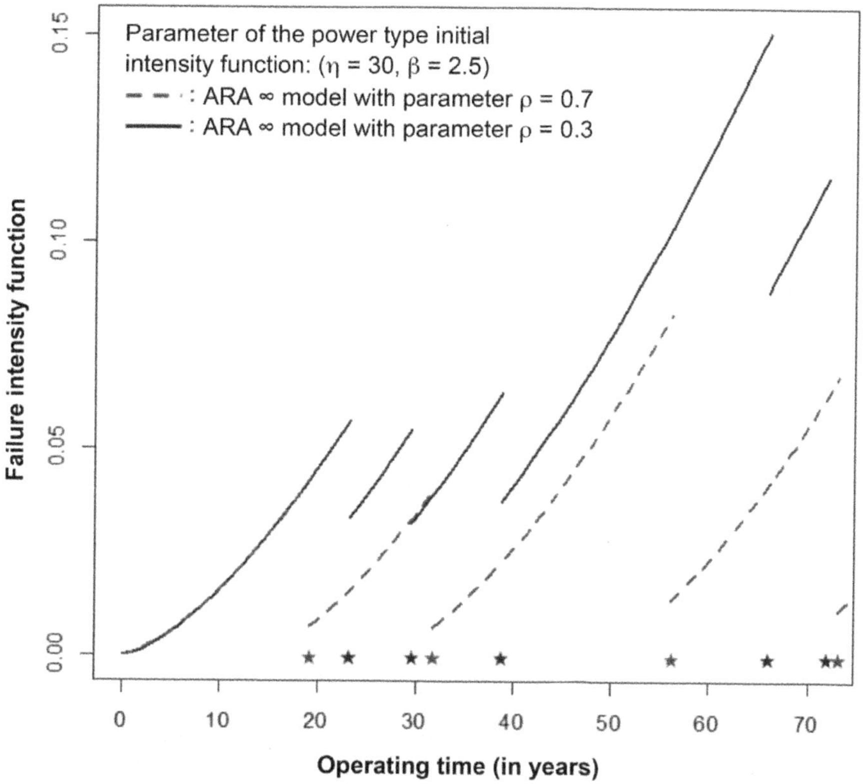

Figure 5.5. *Failure intensity functions for two ARA∞ models with maintenance efficiency parameter $\rho = 0.7$ and $\rho = 0.3$, with the same power type initial intensity function with parameters ($\eta = 30\ years\ ; \beta = 2.5$). The stars on the x-axis correspond to the failure times simulated with the two models. The intensity function presents a decrease after each corrective maintenance action, the level of which depends on the maintenance efficiency parameter ρ (the higher this parameter, the more efficient the maintenance and the more the failure intensity function is reduced after a maintenance). The graph was generated using the "VAM" package from R software environment. For a color version of this figure, see www.iste.co.uk/lannoy/nuclear.zip*

There are clearly a number of other imperfect maintenance models based on the notion of virtual age or on different concepts: we can for example cite arithmetic or geometric reduction of intensity models or quasi-renewal processes.

We will now give some ideas about the case of a piece of equipment on which both corrective and preventive maintenance actions are carried out. We recall that among preventive maintenance actions, the standard [ASS 18] distinguishes, on the one hand, the predetermined maintenance, which is carried out in accordance with established intervals of time or number of units of use but without previous condition investigation, and, on the other hand, conditional maintenance that includes assessment of physical conditions, analysis and the possible ensuing maintenance actions. From the mathematical modeling perspective, what distinguishes conditional maintenance is the fact that, after one maintenance is carried out, the time of the next preventive maintenance action is unknown: it will be determined according to future observations of monitoring carried out on the piece of equipment (not yet known at the moment of the last maintenance). Thus, conditionally on all the observations made at the time of the last maintenance, the date of the next preventive conditional maintenance is random. This is not the case for predetermined maintenance, in particular for maintenance following a pre-established maintenance schedule before the piece of equipment is out into service which we will consider next. To be able to model the two corrective and predetermined preventive maintenance processes, it is sufficient to choose, then to combine together, an initial intensity function $h(.)$ (that characterizes the intrinsic behavior of the new device in the absence of maintenance), a maintenance effect model for corrective maintenance (e.g. AGAN, ABAO, ARA, Brown–Proschan, and so on) and a maintenance effect model for preventive maintenance (e.g. AGAN, ABAO, ARA, Brown–Proschan, and so on), given that by default, there is no particular motivation to suppose that corrective and preventive maintenances present the same effect on the reliability of the piece of equipment (which would then lead us to characterize them by the same model). Applications of this type of models (with corrective and predetermined preventive maintenance) to repairable devices can be found in [PRO 11] and [REM 13]. More precisely, this last reference presents the use of some virtual age models for the purpose of optimization, on the basis of both technical and economic considerations, of the periodicity of preventive maintenance carried out on the boiler of a thermal power station, perfectly illustrating the character of decision support tools for these models.

We will conclude this section by specifying that modeling conditional preventive maintenance or the consideration of influencing factors (either on the intrinsic behavior of the device, or on the effects of maintenance actions) are currently the subject of research to make further progress in the modeling of repairable devices.

5.4.5. *Stochastic degradation models*

The state of health of a passive (structural) component undergoing a degradation mechanism is generally monitored with the help of inspections during which information on its level of degradation $D_t, t \geq 0$, is collected. Suppose to begin with that inspections carried out allow a quantitative estimate of the degradation, such as measurements on the depth of stress corrosion (SC) cracks. We also make the hypothesis that OF data are made of measured values $D_{i,j}$ of the degradation indicator, where index i characterizes the component, $1 \leq i \leq n$, n being the number of components monitored, and where index j defines the inspection carried out at time $t_{i,j}$ on the component i, with $1 \leq j \leq m_i$, m_i being the total number of visits carried out on the component i. We therefore have for each component repeated measurements representative of how its degradation level has evolved. Two main families of models are proposed in the literature to represent these types of data [KAH 16].

The first one is made of models known as degradation path models [MEE 98]. These models aim to represent correctly all the degradation paths observed. To do this, they suppose that the degradation phenomenon is deterministic, even though there may be a variability proper to each component (due to the properties of the device or to different manufacturing conditions or usage – see section 5.4.3), then translated by the introduction of random effects. We can thus write:

$$D_{i,j} = \mathcal{D}(t_{i,j}, \beta_{1,i}, \dots, \beta_{k,i}) + \varepsilon_{i,j}, 1 \leq i \leq n, 1 \leq j \leq m_i \qquad [5.14]$$

where $\mathcal{D}(.)$ is a deterministic function that represents the true degradation dynamic, $\varepsilon_{i,j}$ are distributed following a centered Gaussian distribution $\mathcal{N}(0, \sigma_\varepsilon^2)$ with variance σ_ε^2 and $\beta_{l,i}$, $1 \leq l \leq k$, are random parameters that can differ from one component to another or may be common to many of them. Conditionally on the values taken by these parameters $\beta_{l,i}$ for component i, chance is entirely conveyed by the $\varepsilon_{i,j}$, which can then be interpreted as residual deviations from the true level of degradation $\mathcal{D}(.)$, that is, as measurement errors. Conceptually, degradation path models are very similar to generalized nonlinear models (of which the standard linear regression model is a particular case) and they are particularly well adapted when we have a deterministic description of the underlying phenomenon (relying for example on physical considerations) making it possible to help to specify the shape of the $\mathcal{D}(.)$ function (linear, convex, concave, etc.). Moreover, in some simple cases, we can characterize explicitly the distribution of the random operating duration T_ρ needed for the degradation level to reach a fixed threshold denoted ρ. Mathematically, T_ρ is expressed as $T_\rho = \inf_{t \geq 0}\{t | D_t \geq \rho\}$ and can be interpreted as the lifetime of the passive component if exceeding the level of degradation ρ defines the failure criteria (see section 5.2). Meeker and Escobar [MEE 98] give the example where the $\mathcal{D}(.)$ function is linear $\mathcal{D}(t_{i,j}) = \beta_1 + \beta_{2,i} t_{i,j}$, with β_1 a constant parameter common to all the components and $\beta_{2,i}$ a random effect specific to each component i supposed to be log-normal distributed: it can be shown in this case that lifetime T_ρ follows (also) a log-normal distribution, whose parameters are known. Another way of approximating the distribution of T_ρ numerically is to proceed using Monte Carlo simulations [HON 15]. The simple example above shows in any case one of the limits of this family of models: if the random effects $\beta_{l,i}$ are known (and the measurement errors $\varepsilon_{i,j}$ neglected), the component's lifetime is deterministic; this can lack physical realism.

The second category of models relies on the idea that degradation measurements are realizations of a stochastic process, generally supposed to belong to the family of Lévy processes, of which Wiener, gamma and inverse Gaussian processes are the three most widespread representatives for modeling degradation phenomena. A Lévy process [KAH 16] is a stochastic process that verifies the following mathematical properties:

– it is a continuous-time process, right-continuous and allowing a limit on the left, that is, the function $t \to D_t, \forall t \geq 0$, is a right-continuous function with left limits (almost certainly);

– it starts at 0, that is, $D_0 = 0$ (almost certainly);

– its increments are independent, that is, the random variables $D_{t_1}, D_{t_2} - D_{t_1}, \dots, D_{t_n} - D_{t_{n-1}}$ (representing successive degradation increments in our case) are independent $\forall\ 0 \leq t_1 \leq t_2 \leq \dots \leq t_n$;

– its increments are stationary, that is, the random variables $D_t - D_s$ and D_{t-s} have the same probability distribution $\forall\ 0 \leq s < t$.

The idea underlying the use of the family of Lévy processes is that the level of degradation reached at the end of an operating duration is the result of an accumulation, over disjoint time intervals, of small, independent degradation increments. It must be emphasized that members of the family of Lévy processes all meet the Markov property, according to which the next degradation increment depends only on the current level of deterioration: all dependence on the process's past is summarized in the present state, which, in some examples, can constitute an unrealistic hypothesis compared to the physics of the phenomenon.

To present the Wiener process, we first introduce the Brownian motion, denoted as B_t. This is a Lévy process for which B_t follows a centered Gaussian distribution $\mathcal{N}(0, t)$ with the variance $t, \forall t > 0$. We then define the Wiener process with linear drift with parameters $\boldsymbol{\theta} = (\mu, \sigma), \sigma > 0$, by $D_t = \mu t + \sigma B_t, \forall t \geq 0$, with B_t a Brownian motion, μ a real drift parameter (generally positive in practice, whose unit is the ratio of the unit of the degradation indicator and the unit of time) and $\sigma > 0$ a volatility parameter (or dispersion parameter, of the same unit as parameter μ). The degradation increment $D_t - D_s$ between the two instants s and t, $\forall 0 < s < t$, follows a Gaussian distribution $\mathcal{N}\big(\mu(t - s), \sigma^2(t - s)\big)$, with an expected value and variance both directly proportional to the width of the time interval $t - s$ separating the two degradation measurements. This process has the property of independent and stationary increments. Because these increments follow a Gaussian distribution, the Wiener process with linear drift can present positive or negative increments between two consecutive instants (see Figure 5.6). In the context of degradation modeling, this characteristic can prove relevant for describing the behavior of a component displaying a self-repair capacity or one exposed to fluctuating load levels (resulting in the closure of a crack for example). This stochastic process is also adapted in the presence of measurement uncertainties or noisy signals, which can lead to degradation increments measured as negative between two times, although the underlying physical degradation mechanism is necessarily increasing. One interesting property of the Wiener process with linear drift lies in knowledge of the distribution of the operating time T_ρ needed to reach a fixed degradation threshold ρ. In fact, it can be shown [KAH 16] that T_ρ follows an inverse Gaussian distribution $IG\left(\frac{\rho}{\mu}, \frac{\rho^2}{\sigma^2}\right)$ with expectation $\frac{\rho}{\mu}$ and shape parameter $\frac{\rho^2}{\sigma^2} > 0$, the probability density function of which is given by $f_{T_\rho,IG}(t|\mu,\sigma) = \frac{\rho}{\sqrt{2\pi\sigma^2 t^3}} \exp\left(-\frac{(\rho-\mu t)^2}{2\sigma^2 t}\right), \forall t \geq 0$, and with variance $\frac{\rho\sigma^2}{\mu^3}$. The Wiener process with linear drift can be extended in different ways, for example:

– by taking account of initial, non-zero level of degradation d_0 and an operating time $t_0 > 0$ from which the accumulation of degradation begins: $D_t = d_0 + D_{t-t_0} = d_0 + \mu(t - t_0) + \sigma B_{t-t_0}, \forall t \geq t_0$;

– by considering a nonlinear drift $\xi(t)$, where $\xi(.)$ is an increasing function, continuous and verifying conditions $\xi(0) = 0$ and $\lim_{t\to+\infty} \xi(t) = +\infty$ (e.g. a power function $\xi(t) = t^\gamma, \gamma > 0$): $D_t = \mu\xi(t) + \sigma B_t, \forall t \geq 0$. In this case, the degradation increments are still independent, but they are no longer stationary: this process is not therefore part of the family of Lévy processes;

– by operating a complete transformation of the time scale, using, as for the Wiener process with nonlinear drift, a function $\xi(.)$ so that: $D_t = \mu\xi(t) + \sigma B_{\xi(t)}, \forall t \geq 0$. There again, the degradation increments are independent but non-stationary.

With the Wiener process, the gamma process is the most widespread in the field of degradation modeling [VAN 09]. The homogeneous (or stationary) gamma process with shape parameters $a > 0$ (of a unit inverse to that of time) and rate $b > 0$ (of a unit inverse to that of the degradation indicator) is time-continuous, right-continuous and allowing a left limit, which starts from 0, with independent stationary increments $D_t - D_s, \forall\, 0 \leq s < t$, following gamma distribution $\mathcal{G}(a(t - s), b)$ with shape parameter $a(t - s)$ and rate parameter b. Recall that a positive random variable D follows a gamma distribution $\mathcal{G}(a, b)$ with parameters $\boldsymbol{\theta} = (a, b)$ if its probability density function is written as $f_{D,\mathcal{G}}(d|\,\boldsymbol{\theta}) = \frac{b^a}{\Gamma(a)} d^{a-1} \exp(-bd), \forall d \geq 0$, with $\Gamma(a) = \int_0^{+\infty} x^{a-1} \exp(-x)dx$ the Euler gamma function. Consequently, the degradation level D_t at the end of the operating time $t > 0$ follows a gamma distribution with shape parameter at and rate parameter b, whose expected value is $\mathbb{E}(D_t) = \frac{at}{b}$ and whose variance is $\mathbb{V}(D_t) = \frac{at}{b^2}, \forall t > 0$: the homogeneous gamma process, which is part of the family of Lévy process, is therefore a non-decreasing process (the degradation cannot decrease with time) and with an expectation and variance that are linear following the component's operating duration (see Figure 5.6). A singular property of the gamma process is that it can be seen as the limit of a particular compound Poisson process [KAH 16], the general shape of which is given by $\sum_{i=1}^{N_t} D_i$, where the random cumulative number N_t of jumps made over the operating duration $t \geq 0$ follows a Poisson distribution and the random sizes of the jumps $D_i, i \geq 2$, are identically distributed and independent from one another and from the counting process N_t. This mathematical property joins the physical sense according to which some degradation phenomena can result from external shocks, which will occur randomly over time (following a Poisson process) with shock magnitudes that are also random (following a gamma distributed random variable). All these characteristics explain the popularity of the gamma process for modeling cumulative physical degradation phenomena, such as fatigue or corrosion, which are often encountered in practice.

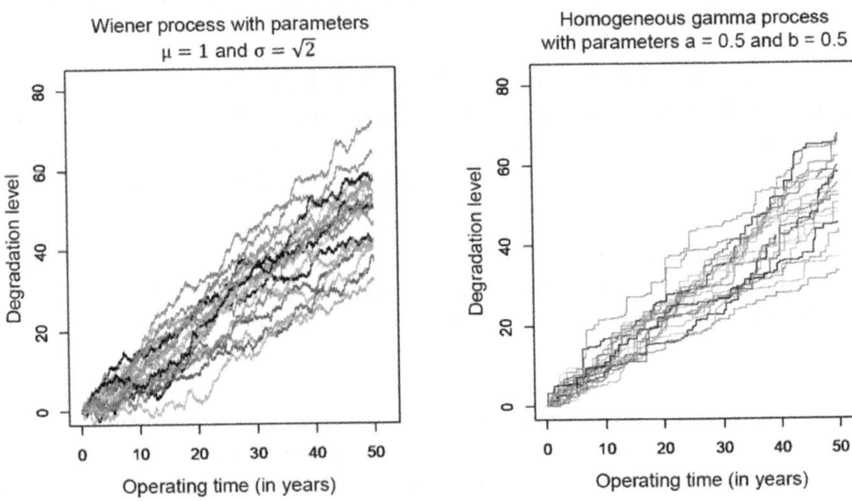

Figure 5.6. *On the left: 20 degradation paths simulated according to a Wiener process with parameters* $(\mu = 1, \sigma = \sqrt{2})$. *On the right: 20 degradation paths simulated according to a homogeneous gamma process with parameters* $(a = 0.5, b = 0.5)$, *with the same expected value and variance as the Wiener process. The degradation increments are not strictly positive for the Wiener process, unlike the gamma process. The intrinsic variability between paths simulated with the same process is clearly illustrated, just like the average linear evolution drift. This graph was generated with the "yuima" package from R software environment. For a color version of this figure, see www.iste.co.uk/lannoy/nuclear.zip*

Even if it is possible to determine easily the distribution of the random operating duration T_ρ needed to reach a fixed degradation threshold ρ, it is difficult in practice to manipulate it for numerical reasons and it is preferable to approximate it using the Birnbaum–Saunders distribution (see [PAR 14a]). The gamma process also presents the advantage of making it possible to evaluate (fairly easily) the distribution of the component's remaining useful life, that is, the (random) operating duration needed to reach the degradation threshold ρ given the level of degradation d_u at time $u \geq 0$ (in other words the delay to pass from the degradation level d_u measured at time u at limit threshold ρ).

The stationary gamma process can be rendered non-homogeneous by applying a complete transformation of the time scale, via a continuous, increasing nonlinear function $\xi(.)$, verifying $\xi(0) = 0$ and $\lim_{t\to+\infty} \xi(t) = +\infty$, for example a power type function $t^c, c > 0$, so that $D_t - D_s$ then follows a gamma distribution $\mathcal{G}(a(t^c - s^c), b)$ with parameters $\boldsymbol{\theta} = (a, c, b)$. This extension makes the degradation increments non-stationary (the non-homogeneous gamma process is not therefore part of the family of Lévy processes), but makes it possible to model the average degradation dynamics which do not evolve linearly with time, since we then have $\mathbb{E}(D_t) = \frac{a\xi(t)}{b}$. Many other extensions of the gamma process are present in the literature, such as the extended gamma process [ALM 17] making it possible to free ourselves from the constant expectation/variance ratio, the integration of influencing factors or random effects (see section 5.4.3), consideration of the measurement uncertainty on the level of degradation observed or the addition of a random initiation duration of the physical deterioration phenomena [BOR 16]. This last reference presents an application of the model developed to a family of EDF nuclear power plant components, for which we seek to evaluate as precisely as possible the remaining useful life. On these components, two inspection procedures are used during a preventive visit: the first makes it possible to know the number of flaws that have developed on each component, whereas the second gives information on the size of the largest indication present during the inspection (we do not therefore have measurements of other flaws that may exist). A hybrid model, pairing the stochastic Poisson type counting process (see section 5.4.4) for the initiation of degradations and a stochastic gamma process for the evolution of their size in time, has been proposed to take advantage of all the information available and to consider the two stages of initiation and propagation of deterioration.

We conclude this presentation of current continuous stochastic processes for degradation modeling by mentioning the homogeneous inverse Gaussian process with parameters $(\mu, \eta), \mu > 0, \eta > 0$ [WAS 68]. This is a Lévy process for which the level of degradation D_t at the end of the operating duration $t > 0$ follows an inverse Gaussian distribution $IG(\mu t, \eta t^2)$ with expectation $\mathbb{E}(D_t) = \mu t$ and shape parameter ηt^2, the variance of which is given by $\mathbb{V}(D_t) = \frac{\mu^3 t}{\eta}$. Just as for the gamma process, the deterioration increments are strictly positive and allow modeling of cumulative degradation. This process, which has been the subject of many developments, remains generally less popular than the gamma process, particularly today because it lacks physical interpretability [YE 15].

To finish this overview of stochastic degradation processes, it makes sense to mention very briefly the case where visits carried out on the passive component provide a qualitative estimate of the deterioration level, such as a visual inspection record specifying a "healthy state", "degraded state", "much degraded state" or a "failure state", without giving a measurement of the damage. The random continuous processes presented previously are no longer therefore adapted and other stochastic processes must be used, called "discrete state (space) models" (also called multi-state models), of which the Markov process is the best-known representative. A state-transition diagram is often used to represent these processes (see Figure 5.7): each node in this graph represents a possible state of degradation and the edges (or links) characterize the transitions that exist between the nodes. The evolution dynamic of D_t is then governed by the paths possible between the states and by the probabilities (in discrete-time) or the intensities (in continuous-time) of transition between states.

When the information useful for future degradation prediction is contained (summarized) entirely in the present state of the process and is not dependent on earlier states, it is called a Markov process, which is the subject of standard [INT 06b] and for which you can find in [PAR 14b] an application to a family of nuclear power plant components (in this industrial example based on real OF data, the probabilities of transition from one state of degradation to the other are modeled according to several influencing factors, see section 5.4.3). When this hypothesis is not realistic for the component studied and the transition between present and future states depends on factors other than the current state (such as for example the past states, the component's operating duration or the time the component has spent in its current state), the Markov process is no longer relevant and more elaborate models must be adopted, such as for example semi-Markov models or piecewise deterministic Markov processes.

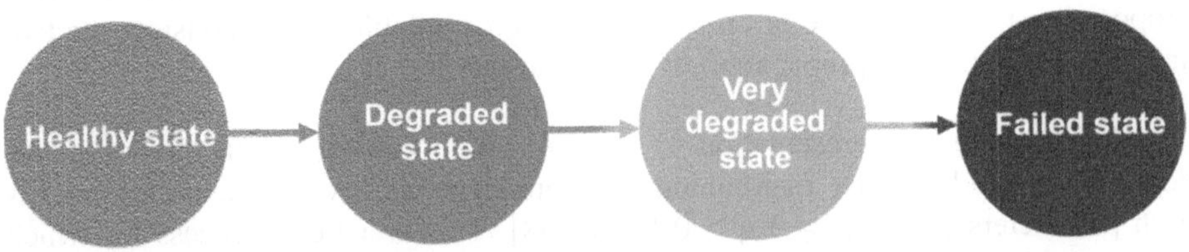

Figure 5.7. *Example of a transition diagram with four degradation states. For a color version of this figure, see www.iste.co.uk/lannoy/nuclear.zip*

We will conclude this section by specifying that multivariate degradation (when several degradation mechanisms are present and measured on the component), the consideration of influencing factors [BAG 01] or the integration of maintenance impact on the degradation level are currently the subject of research to develop degradation modeling.

5.5. Frequentist inference

Once the work of modeling is accomplished and the parametric probabilistic model established, it makes sense to estimate its d parameters $\boldsymbol{\theta} = (\theta_1, \ldots, \theta_d)$. The frequentist inference approach relies exclusively on the distribution of the available OF data (unlike Bayesian approach, see section 5.6) and relies on the paradigm according to which parameters $\boldsymbol{\theta}$ are unknown but deterministic, that is, that they take fixed values, to be determined (approximated) so that the selected model, written $\mathcal{M}(\boldsymbol{\theta})$, fits all the data as well as possible. It will subsequently be assumed that $\boldsymbol{\theta}$ belongs to a set $\Theta \subset \mathbb{R}^d$: for example, in the case of the Weibull model $\mathcal{W}(\eta, \beta)$ presented in section 5.4.2, we have $\mathcal{M}(\boldsymbol{\theta})$ the family of Weibull distributions $\mathcal{W}(\eta, \beta)$, $\boldsymbol{\theta} = (\eta, \beta)$, $d = 2$ and $\Theta = \mathbb{R}_+^* \times \mathbb{R}_+^*$, the two constants η and β being strictly positive. Scientific literature abounds with frequentist techniques allowing this statistical inference work to be carried out, among them the method of moments and the least squares method. We will be content in this chapter to introduce the maximum likelihood method, which presents several advantages (it can be applied to a very broad variety of models and data – continuous, discrete, categorical, censored, truncated etc.; furthermore, it presents good mathematical properties making it possible to guarantee, under some regularity conditions for the model[3], that the estimations it provides are, asymptotically [i.e. for a large number of failures observed], more "precise" [i.e. with weaker variance] than those of competing methods), which we will not discuss in more detail here.

The idea of the maximum likelihood method is to retain for parameter values $\boldsymbol{\theta}$ those that maximize the probability of observing the data manipulated with the model $\mathcal{M}(\boldsymbol{\theta})$ considered. Indeed, intuitively, a combination of values of the d parameters associated with a high probability will, in view of the data and the model, be preferable (since it is more plausible, probable or likely) to another combination associated with a lower probability. The method relies on the likelihood function, denoted $\mathcal{L}_{\mathcal{M}}(\text{data}|\boldsymbol{\theta})$, which is equal (or approximately proportional) to the probability of observing the available data with the model $\mathcal{M}(\boldsymbol{\theta})$ chosen to represent them. In the frequentist approach, all the information provided by the data

3 Such as the fact that (the log of) the probability density function is twice continuously differentiable on Θ and that the support of the probability density function does not depend on the θ parameters.

on $\boldsymbol{\theta}$ is contained entirely in this likelihood function, and for a fixed OF dataset and model, the likelihood function can be seen as a function of the unknown $\boldsymbol{\theta}$ parameters. The expression of this function depends both on the type of data manipulated and on the model chosen, the great diversity of which we have glimpsed in sections 5.2 and 5.4: writing of $\mathcal{L}_{\mathcal{M}}(.|.)$ must therefore by adapted for each case encountered, which we will illustrate in the rest of this section.

Let us suppose to begin with that we seek to characterize the probability distribution of the lifetime T of a family of non-repairable active components, for which we have an OF composed of all the types of censoring below (see section 2.6.2 of Chapter 2):

– exact (or complete) failure data, linked to the components that have broken down exactly at the end of the known operating durations $t_i, i \geq 1$ (the t_i are therefore realizations of random variable T);

– right-censored data (or survival data), linked to components known to be in good working order at the end of their operating durations $t_j^d, j \geq 1$ (for these components, failures occur beyond t_j^d, that is, $T > t_j^d$, the associated probability being expressed by $\mathbb{P}(T > t_j^d)$);

– left-censored data, linked to components known only to have broken down before reaching time $t_k^g, k \geq 1$ (failure therefore occurred before t_k^g, i.e. $T < t_k^g$, the associated probability being expressed by $\mathbb{P}(T < t_k^g)$);

– interval-censored data, linked to components known simply to have broken down between operating durations t_h^{id} and t_h^{ig}, $h \geq 1$ (i.e. $t_h^{id} < T < t_h^{ig}$, the associated probability is expressed by $\mathbb{P}(t_h^{id} < T < t_h^{ig}) = \mathbb{P}(T < t_h^{ig}) - \mathbb{P}(T < t_h^{id})$).

We also hypothesize that the $\mathcal{M}(\boldsymbol{\theta})$ model is retained to represent the data (for example the Weibull distribution $\mathcal{W}(\eta, \beta)$), for which the probability density function will be denoted $f_{T,\mathcal{M}}(.|\boldsymbol{\theta})$, the cumulative distribution function $F_{T,\mathcal{M}}(.|\boldsymbol{\theta})$ and the survival function $S_{T,\mathcal{M}}(.|\boldsymbol{\theta})$. If the data are independent (i.e. if the occurrence of one has not modified the probabilities of occurrence of the others – this is generally the case in practice in the absence of common mode or common cause failure) and identically distributed (i.e. all coming from the same underlying random mechanism represented by $\mathcal{M}(\boldsymbol{\theta})$), the likelihood function is then expressed as the product of the individual probabilities linked to each piece of data, each of them being expressed differently depending on whether the data is exact, right-censored, left-censored or interval-censored [MEE 98]:

$$\mathcal{L}_{\mathcal{M}}(\text{data}|\boldsymbol{\theta}) = \prod_{i\geq 1} f_{T,\mathcal{M}}(t_i|\boldsymbol{\theta}) \prod_{j\geq 1} S_{T,\mathcal{M}}(t_j^d|\boldsymbol{\theta}) \prod_{k\geq 1} F_{T,\mathcal{M}}(t_k^g|\boldsymbol{\theta}) \prod_{h\geq 1} \left(F_{T,\mathcal{M}}(t_h^{ig}|\boldsymbol{\theta}) - F_{T,\mathcal{M}}(t_h^{id}|\boldsymbol{\theta})\right). \quad [5.15]$$

The maximum likelihood method then comes down to estimating $\boldsymbol{\theta}$ by $\widehat{\boldsymbol{\theta}}$ verifying: $\widehat{\boldsymbol{\theta}} = \text{argmax}_{\boldsymbol{\theta}\in\Theta} \mathcal{L}_{\mathcal{M}}(\text{data}|\boldsymbol{\theta})$. To illustrate the approach, we take the example of an OF dataset coming from observation of n components, formed (only) of r complete failure data at times $t_i, 1 \leq i \leq r$, and of $n - r$ right-censored data $t_j^d, 1 \leq j \leq n - r$, supposed to be independent and identically distributed following the exponential $\mathcal{E}(\lambda)$ model with rate parameter $\lambda > 0$ (recall that this is a Weibull distribution $\mathcal{W}\left(\eta = \frac{1}{\lambda}, \beta = 1\right)$, see section 5.4.2 – its instantaneous failure rate function is constant over time $h_{T,\mathcal{E}}(t|\lambda) = \lambda, \forall t \geq 0$, and its probability density and survival functions are expressed, respectively, by $f_{T,\mathcal{E}}(t|\lambda) = \lambda\exp(-\lambda t), \forall t \geq 0$ and $S_{T,\mathcal{E}}(t|\lambda) = \exp(-\lambda t), \forall t \geq 0$). In this particular case, the likelihood function is written as:

$$\mathcal{L}_{\mathcal{M}}(\text{data}|\theta) = \mathcal{L}_{\mathcal{E}}\left((t_i)_{1\leq i\leq r}, \left(t_j^d\right)_{1\leq j\leq n-r} |\lambda\right) = \prod_{i=1}^{r} f_{T,\mathcal{E}}(t_i|\lambda) \prod_{j=1}^{n-r} S_{T,\mathcal{E}}\left(t_j^d|\lambda\right) = \prod_{i=1}^{r} \lambda\exp(-\lambda t_i) \prod_{j=1}^{n-r} \exp\left(-\lambda t_j^d\right) = \lambda^r\exp\left(-\lambda\left(\sum_{i=1}^{r} t_i + \sum_{j=1}^{n-r} t_j^d\right)\right) \quad [5.16]$$

Determining the maximum likelihood estimator of parameter λ leads to finding the maximum of the likelihood function $\lambda \to \mathcal{L}_{\mathcal{E}}(\text{data}|\lambda)$ or, in strictly equivalent fashion, the log of this function, called log-likelihood, $\lambda \to \ln\left(\mathcal{L}_{\mathcal{E}}(\text{data}|\lambda)\right)$ (since the log function is strictly monotonous, it reaches its maximum in the same place[4] as the $\mathcal{L}_{\mathcal{E}}(\text{data}|\lambda)$ function). To do this, we seek the value $\hat{\lambda}$ of $\lambda > 0$ which cancels the (first) derivative of the log-likelihood:

$$\frac{d}{d\lambda}\ln\left(\mathcal{L}_{\mathcal{E}}(\text{data}|\lambda)\right) = \frac{d}{d\lambda}\left(r\ln(\lambda) - \lambda\left(\sum_{i=1}^{r} t_i + \sum_{j=1}^{n-r} t_j^d\right)\right) = \frac{r}{\lambda} - \left(\sum_{i=1}^{r} t_i + \sum_{j=1}^{n-r} t_j^d\right) \quad [5.17]$$

that is, $\hat{\lambda} = \frac{r}{\sum_{i=1}^{r} t_i + \sum_{j=1}^{n-r} t_j^d}$ (which is indeed the value at which $\mathcal{L}_{\mathcal{E}}(\text{data}|\lambda)$ takes its maximum, because the second derivative is strictly negative: $\frac{d^2}{d\lambda^2}\ln\left(\mathcal{L}_{\mathcal{E}}(\text{data}|\lambda)\right) = -\frac{r}{\lambda^2} < 0, \forall\lambda > 0$). We find the formula currently used (sometimes wrongly, since it is only valid under the hypothesis of an exponential distribution), which consists of estimating a failure rate (function) by the ratio of the observed number of failures, r, and the total operating duration of the n components (total time on test, TTT), $\sum_{i=1}^{r} t_i + \sum_{j=1}^{n-r} t_j^d$, whether they have or have not broken down (moreover, the

4 If this maximum even exists.

importance can easily be seen here of not omitting positive OF made of $n-r$ survival data, on pain of over-estimating the value of λ).

Let us illustrate a second example, supposing now that we wish to model the lifetime T of non-repairable components with the help of a Cox model to take account of influencing factors (see section 5.4.3). Take the example of an OF dataset taken from that observation of n components, made up of r exact failure data at the end of the operating durations (without tied values) $t_i, 1 \leq i \leq r$, and $n-r$ survival data, supposed to be independent. We write $\mathbf{Z}_i, 1 \leq i \leq n$, the vector of the values of the p covariates for component i, $\delta_i, 1 \leq i \leq n$, the indicator function which is equal to 1 if component i has failed, 0 otherwise, and R(t) all the components in good working order and not censored just before t (this set is therefore made of the components at risk, which may undergo failure at the end of the operating duration t, given all that has happened before). The (partial) likelihood function is then expressed by [LAW 03]:

$$\mathcal{L}_{\mathcal{M}}(\text{data}|\boldsymbol{\theta}) = \mathcal{L}_{\text{Cox}}((t_i, \delta_i, \mathbf{Z}_i)_{1 \leq i \leq n}|\boldsymbol{\beta}) = \prod_{i=1}^{n} \left(\frac{h_{T,\text{Cox}}(t_i, \mathbf{Z}_i|\boldsymbol{\beta})}{\sum_{j=1}^{n} Y_j(t_i) h_{T,\text{Cox}}(t_i, \mathbf{Z}_j|\boldsymbol{\beta})} \right)^{\delta_i} = \prod_{i=1}^{n} \left(\frac{\exp(\boldsymbol{\beta}^T \mathbf{Z}_i)}{\sum_{j=1}^{n} Y_j(t_i) \exp(\boldsymbol{\beta}^T \mathbf{Z}_j)} \right)^{\delta_i} \quad [5.18]$$

where $\boldsymbol{\beta}$ is the vector of p unknown parameters to be estimated included in the Cox model and $Y_i(t) = \mathbb{I}_{\{t_i \geq 0\}}$, i.e. $Y_i(t) = 1$ if, and only if, $i \in R(t), \forall t \geq 0$.

Let us proceed by considering that we seek to model the process of repeated breakdowns of a repairable piece of equipment on which only corrective maintenance is carried out (see section 5.4.4). Let us take the example of an OF dataset taken from observation of a (single) device over the time interval $[t_0 = 0; t]$ and made up of operating durations t_i, $1 \leq i \leq n$, at the end of which the component has broken down (the t_i correspond, too, to instants of corrective maintenance, under the hypothesis that delays in intervention and the durations of repairs are left out; t is the duration of the end of the component's observation, corresponding to a survival data). If we choose an initial Weibull type intensity function, $h_{T,\mathcal{W}}(t|\eta, \beta) = \frac{\beta}{\eta}\left(\frac{t}{\eta}\right)^{\beta-1}, \eta > 0, \beta > 0, \forall t \geq 0$, to model the intrinsic behavior of the new, unmaintained piece of equipment, and an ARA∞ model with maintenance efficiency parameter $\rho \leq 1$, the likelihood function is then expressed by [COO 07]:

$$\mathcal{L}_{\mathcal{M}}(\text{data}|\boldsymbol{\theta}) = \mathcal{L}_{\text{ARA}\infty}((t_i)_{1 \leq i \leq n}|\eta, \beta, \rho) = \prod_{i=1}^{n} \lambda_{\text{ARA}\infty}(t_i|\eta, \beta, \rho) \exp\left(-\sum_{i=1}^{n+1} \int_{t_{i-1}}^{t_i} \lambda_{\text{ARA}\infty}(u|\eta, \beta, \rho) du\right) \quad [5.19]$$

with $t_{n+1} = t$ and $\lambda_{ARA\infty}(t|\eta, \beta, \rho) = h_{T,\mathcal{W}}\left(t - \rho \sum_{j=0}^{N_{t^-}-1}(1-\rho)^j T_{N_{t^-}-j} | \eta, \beta\right), \forall t \geq 0$, the (failure) intensity function of the ARA∞ model.

Let us finish this panorama of likelihood function expressions by supposing that we wish to model the level of degradation $D_t, t \geq 0$, of a passive (structural) component undergoing a damage mechanism (see section 5.4.5). Take the example of an OF dataset taken from n inspections allowing a quantitative assessment of the degradation of the component studied, made of n degradation levels $d_{t_i}, 1 \leq i \leq n$, measured at operating durations $t_i, 1 \leq i \leq n$. If we choose a homogeneous gamma process with shape parameter $a > 0$ and rate parameter $b > 0$ to model these data, the likelihood function is then expressed by [KAH 16]:

$$\mathcal{L}_{\mathcal{M}}(\text{data}|\boldsymbol{\theta}) = \mathcal{L}_{\text{proc.gamma}}\left(\left(\Delta d_{\Delta t_i}\right)_{1 \leq i \leq n} | a, b\right) = \prod_{i=1}^{n} f_{D,\mathcal{G}}\left(\Delta d_{\Delta t_i} | \, a\Delta t_i, b\right) \quad [5.20]$$

where $f_{D,\mathcal{G}}(.\,|a, b)$ is the probability density function of the gamma distribution $\mathcal{G}(a, b)$ and $\Delta d_{\Delta t_i} = d_{t_i} - d_{t_{i-1}}, 1 \leq i \leq n$, the degradation increments observed over each time interval $\Delta t_i = t_i - t_{i-1}, 1 \leq i \leq n$, between two successive inspections (with $t_0 = d_0 = 0$).

We conclude this introduction to the maximum likelihood estimation method, and more generally to the frequentist inference approach, by specifying the following elements.

– It is not possible to emphasize enough the fact that all frequentist inference techniques rely exclusively on the OF data available: the quality of the result of the reliability study depends therefore on the quality of the data.

– In practice, the OF may not show any failure, typically when we study very reliable components or components broadly designed with safety in mind. Even though the literature proposes solutions to estimate the failure rate parameter of an exponential distribution in this example (see [MEE 98] or [GER 17] for example), it makes sense to enquire into the merits of using the frequentist inference approach, whereas the Bayesian approach (which relies on prior knowledge in addition to data, see section 5.6) or structural reliability analysis (on the basis of physical models or numerical simulation codes, see Chapter 4) seem more able to provide robust response elements in this special case.

– The maximum likelihood method estimator displays the property of being functionally invariant, in the sense that if $\widehat{\boldsymbol{\theta}}$ is the estimator of $\boldsymbol{\theta}$, then the maximum likelihood estimator $\widehat{g(\boldsymbol{\theta})}$ of any function $g(\boldsymbol{\theta})$ is $g(\widehat{\boldsymbol{\theta}})$. This characteristic is very useful since it makes it possible easily, given $\widehat{\boldsymbol{\theta}}$, to estimate any indicator (scalar or functional) of interest dependent on $\boldsymbol{\theta}$ (we then speak of estimation using a plug-in approach, since it is enough to "insert" $\widehat{\boldsymbol{\theta}}$ into $g(.)$ to have its estimator $\widehat{g(.)}$).

– The maximum likelihood optimization problem $\widehat{\boldsymbol{\theta}} = \text{argmax}_{\boldsymbol{\theta}\in\Theta}\, \mathcal{L}_{\mathcal{M}}(\text{data}|\boldsymbol{\theta})$ can require the use of numerical tools to solve it (no explicit analytical solution exists systematically, as is the case for example for Weibull distribution $\mathcal{W}(\eta, \beta)$). It may also have multiple solutions (the maximum likelihood estimator is not therefore unique), or indeed it may not have any at all (the maximum likelihood estimator does not therefore exist, as is the case for example for the three-parameter Weibull model when $\beta \leq 2$ – see [LAW 03]).

– Even though it is useful to have inferred values of the model's parameters, these point estimates may be misleading, since they give no information on the proximity of $\widehat{\boldsymbol{\theta}}$ to $\boldsymbol{\theta}$, that is, on the quality (or precision) of the approximations obtained. Moreover, in some cases, it may be preferable to have an upper or a lower bound of $\boldsymbol{\theta}$ rather than an estimation of its true value (e.g. it may be desirable to ensure that the failure rate parameter $\lambda > 0$ of an exponential distribution is below a given threshold). At this limit, the confidence intervals remedy the point estimations. They make it possible to quantify the degree of precision of the estimations of $\boldsymbol{\theta}$ taken from the OF data available, given that there are two main sources of variation on the data that can cause a lack of precision in the estimation: either an insufficient quantity of data, or a measurement noise on the data. Thus, the confidence intervals integrate only the fluctuations in sampling and do not take account of other sources of error or bias, such as for example the choice (possibly erroneous) of model $\mathcal{M}(\boldsymbol{\theta})$ to represent the data (see section 5.7). The confidence intervals are specified at a certain level (or threshold) of confidence, generally 90% in practice (to be modulated however according to the industrial issues and the amount of data available), which expresses the confidence (and not the probability!) that a particular interval contains the true (unknown) value of $\boldsymbol{\theta}$. This level of confidence reflects the idea that if we could have 100 OF datasets, then 90 of the 100 confidence intervals on $\boldsymbol{\theta}$ that could be built on these 100 datasets would contain the true value of the parameters. The narrower a confidence interval, the "more valuable" the information on the unknown parameters provided by the data (this is particularly true the more numerous the data are) and the more precise their estimations $\widehat{\boldsymbol{\theta}}$. There are three main scenarios for building confidence intervals [MEE 17]. In some configurations (rare in practice), the exact probability distribution of $\widehat{\boldsymbol{\theta}}$ is known ($\widehat{\boldsymbol{\theta}}$ is a random variable, since it is a function of the data, i.e. of random realizations of the stochastic mechanism that generated the OF) and it is possible to build exact confidence intervals for $\boldsymbol{\theta}$. Very often (in particular when the OF dataset contains censored data), we only have available a limiting approximation of the distribution of $\widehat{\boldsymbol{\theta}}$ which relies on asymptotic convergence theorems (i.e. when the observed number of failures tends toward infinity): the confidence interval is then only approximate and its interpretation can only be understood under the hypothesis that there is sufficient data (which may not be the case when we have a small amount of OF data). In [MEE 98], you can find many examples of confidence intervals based on this asymptotic approach. The last possibility consists of using data re-sampling techniques (parametric or non-parametric), called bootstrap techniques [EFR 94]. They consist of three stages:

– first, to replicate randomly many times, for example 500 times, the available OF database, to produce 500 new datasets;

– then to estimate the parameters $\boldsymbol{\theta}$ from each of these 500 simulated datasets;

– finally, to use the information provided by the empirical distribution of the 500 estimations $\widehat{\boldsymbol{\theta}}$ taken from the 500 bootstrap datasets to build a confidence interval on $\boldsymbol{\theta}$.

These re-sampling approaches can prove very effective and provide more robust confidence than those relying on asymptotic approximations. The interested reader can consult [EFR 81] and [MEE 98] for more elements on these techniques, in particular in the context of censored data or small databases.

5.6. Bayesian statistics

Unlike the frequentist approach (see section 5.5), the Bayesian statistics rely on the paradigm according to which the d unknown parameters $\boldsymbol{\theta} = (\theta_1, \dots, \theta_d)$ of the model $\mathcal{M}(\boldsymbol{\theta})$ (supposed to have generated the OF data and which is sometimes called the sampling model) are random variables. These random variables follow a (joint) probability distribution called a *prior* distribution, of which the probability density function will be written $\pi(\boldsymbol{\theta})$. This distribution reflects the *prior* knowledge that we may have on the parameters $\boldsymbol{\theta}$. It generally takes the form of a (hyper)parametric model, in the sense that $\pi(.)$ will be made to depend on the $\boldsymbol{\delta}$ parameters, called hyperparameters: thus $\pi(\boldsymbol{\theta}) = \pi(\boldsymbol{\theta}|\boldsymbol{\delta})$. Defining[5] the *prior* distribution of $\boldsymbol{\theta}$ therefore amounts to giving a parametric expression to the probability density function $\pi(.|\boldsymbol{\delta})$ and calibrating $\boldsymbol{\delta}$ from *prior* knowledge [DIA 18]. This may come from the following:

– Physical or mathematical constraints on $\boldsymbol{\theta}$. We know, for example, that the rate parameter of the exponential model $\mathcal{E}(\lambda)$, currently used to represent the lifetime T of a non-repairable component (see section 5.4.2), is necessarily positive: it is therefore natural in this case to retain as *prior* distribution $\pi(.)$ of $\lambda > 0$ a distribution with support on $\mathbb{R}_+^*$, such as the gamma distribution.

– Information on $\boldsymbol{\theta}$ from field experts. A specialist in the domain of maintenance can, for example, have *prior* knowledge on the efficiency of some repair operations [COR 12], thus making it possible to define a range of possible values of the corrective maintenance efficiency parameter ρ of the virtual age model ARA∞ (see

5 We often use the verb "elicit", from the Latin elicio, since it very often involves characterizing the prior knowledge from experts from whom knowledge must be extracted.

section 5.4.4), typically $0 \le \rho \le 1$ if the maintenance is not judged detrimental by the expert (in this case, we can legitimately consider for $\pi(\rho)$ a probability distribution bounded on the interval [0; 1], such as the beta distribution).

– Old OF data (i.e. older than the observation of the OF data available) or analogous OF (i.e. taken from equipment similar to those of the family studied, but which are used in slightly different conditions, or in other industries, or by other users) the statistical analysis of which makes it possible to target possible values of $\boldsymbol{\theta}$.

The *prior* distribution of $\boldsymbol{\theta}$ being established, the Bayesian approach relies on the updating (or actualization) of this distribution by using the information provided by the available OF data, this information being characterized by the likelihood function $\mathcal{L}_{\mathcal{M}}(\text{data}|\boldsymbol{\theta})$ introduced in section 5.5 in the context of the frequentist approach. Formally, this Bayesian inference relies on Bayes' theorem [BAY 63], which provides the conditional probability distribution of $\boldsymbol{\theta}$ given the data observed, called *posterior* distribution, the probability density function of which is expressed by:

$$\pi(\boldsymbol{\theta}|\text{data}) = \frac{\mathcal{L}_{\mathcal{M}}(\text{data}|\boldsymbol{\theta})\pi(\boldsymbol{\theta})}{\int_{\boldsymbol{\theta}^* \in \Theta} \mathcal{L}_{\mathcal{M}}(\text{data}|\boldsymbol{\theta}^*)\pi(\boldsymbol{\theta}^*)d\boldsymbol{\theta}^*} \quad [5.21]$$

with Θ all the possible values of $\boldsymbol{\theta}$ ($\Theta \subset \mathbb{R}^d$). Bayesian inference therefore amounts to conditioning the knowledge, contained in the *prior* distribution, at the outset by observing the data. It may be noted that the denominator $\int_{\boldsymbol{\theta}^* \in \Theta} \mathcal{L}_{\mathcal{M}}(\text{data}|\boldsymbol{\theta}^*)\pi(\boldsymbol{\theta}^*)d\boldsymbol{\theta}^*$ involved in the expression $\pi(\boldsymbol{\theta}|\text{data})$ does not depend on $\boldsymbol{\theta}$: it only plays a normalizing role and the *posterior* probability density function is directly proportional to the product of the likelihood function and of the *prior* probability density function of $\boldsymbol{\theta}$. We are therefore limited, often, to writing $\pi(\boldsymbol{\theta}|\text{data}) \propto \mathcal{L}_{\mathcal{M}}(\text{data}|\boldsymbol{\theta})\pi(\boldsymbol{\theta})$.

To illustrate this approach, we consider the case (already tackled in the context of the frequentist approach, see section 5.5) where we seek to characterize the probability distribution of the lifetime T of a family of active, non-repairable components, for which we have a dataset of OF taken from the observation of n components, made of r exact failure data at the end of the operating durations $t_i, 1 \le i \le r$, and of $n - r$ right-censored data $t_j^d, 1 \le j \le n - r$, supposed to be independent and identically distributed following the exponential $\mathcal{E}(\lambda)$ model with rate parameter $\boldsymbol{\theta} = \lambda > 0$. In this case, the likelihood function is written as $\mathcal{L}_{\mathcal{E}}(\text{data}|\lambda) = \lambda^r \exp\left(-\lambda\left(\sum_{i=1}^{r} t_i + \sum_{j=1}^{n-r} t_j^d\right)\right)$. In the Bayesian paradigm, λ is a strictly positive random variable, to which must be linked a *prior* distribution. We take, for example, a gamma distribution with parameters $\boldsymbol{\delta} = (a, b), a > 0, b > 0$ (which is relevant, because this distribution is supported by real positive numbers):

we then have $\pi(\lambda) = \pi(\lambda|\boldsymbol{\delta}) = \frac{b^a}{\Gamma(a)}\lambda^{a-1}\exp(-b\lambda), \forall\lambda > 0$ by definition of the probability density function of the gamma distribution $\mathcal{G}(a,b)$. Bayesian inference consists of establishing (by combining *prior* knowledge and data) the *posterior* distribution of λ, the probability density function $\pi(\lambda|\text{data})$ which we know is proportional to $\mathcal{L}_{\mathcal{E}}(\text{data}|\lambda)\pi(\lambda)$. In this precise case, this product between likelihood and *prior* density functions is equal to:

$$\lambda^r\exp\left(-\lambda\left(\sum_{i=1}^r t_i + \sum_{j=1}^{n-r} t_j^d\right)\right)\frac{b^a}{\Gamma(a)}\lambda^{a-1}\exp(-b\lambda) = \frac{b^a}{\Gamma(a)}\lambda^{r+a-1}\exp\left(-\lambda\left(\sum_{i=1}^r t_i + \sum_{j=1}^{n-r} t_j^d + b\right)\right). \quad [5.22]$$

Thus, $\pi(\lambda|\text{data}) \propto \lambda^{r+a-1}\exp\left(-\lambda\left(\sum_{i=1}^r t_i + \sum_{j=1}^{n-r} t_j^d + b\right)\right)$: in the term on the right, we may recognize the expression of the probability density function of a gamma distribution $\mathcal{G}\left(r + a, \sum_{i=1}^r t_i + \sum_{j=1}^{n-r} t_j^d + b\right)$. Thus, the *posterior* distribution of the rate parameter $\lambda > 0$ of an exponential model $\mathcal{E}(\lambda)$ under the hypothesis of a *prior* gamma distribution is also a gamma distribution: *prior* and *posterior* distributions belong to the same distribution family and the *prior* gamma distribution is said to be "conjugated" with the exponential model (supposed to have generated the OF data).

This property of conjugation belongs only to some particular combinations of $\pi(\boldsymbol{\theta})$ and $\mathcal{L}_{\mathcal{M}}(\text{data}|\boldsymbol{\theta})$ and, in general, the *posterior* probability density function $\pi(\boldsymbol{\theta}|\text{data})$ has no explicit analytical expression (because of the complexity of establishing the normalizing constant $\int_{\boldsymbol{\theta}^*\in\Theta}\mathcal{L}_{\mathcal{M}}(\text{data}|\boldsymbol{\theta}^*)\pi(\boldsymbol{\theta}^*)d\boldsymbol{\theta}^*$ at the denominator). We must then have recourse to numerical integration methods or Monte Carlo simulation (e.g. through importance sampling or Markov chains), which will make it possible to simulate realizations of $\boldsymbol{\theta}$ numerically with distribution (asymptotically) similar to the *posterior* distribution sought (known almost to a constant).

To conclude this (very brief) introduction to the Bayesian statistics, it seems important to us to specify the following elements.

– In terms of results, the frequentist and Bayesian approaches meet in some borderline cases such as when the amount of OF data is substantial (the information they provide then takes precedence over *prior* knowledge) or when the *prior* distribution is vague (diffuse or uninformative – it provides no information in addition to that already included in the data and the *posterior* distribution is proportional to the likelihood function [MEE 98]).

– By probabilizing the parameter $\boldsymbol{\theta}$ of the sampling model $\mathcal{M}(\boldsymbol{\theta})$ that has generated the data and by introducing *prior* distribution, the Bayesian approach provides more flexibility than the frequentist framework, by making it possible to consider an epistemic uncertainty on $\boldsymbol{\theta}$ and incorporating knowledge external to the data, which can prove useful (indeed necessary) in the domain of the reliability of nuclear power plant equipment, where OF data may not be very informative (very highly censored OF data or OF of limited size, insufficient for estimating $\boldsymbol{\theta}$ precisely). This is therefore an advantage of the Bayesian approach. Conversely, it allocates a fundamental role to *prior* distribution, since it influences directly the result of the Bayesian inference. Indeed, this choice may introduce an intrinsic subjectivity into the analysis, which may be lamented. Eliciting the *prior* distribution should therefore be the subject of quite particular attention [DIA 18] and sensitivity studies on the choice of $\pi(\boldsymbol{\theta})$ should be carried out to justify the robustness of the Bayesian inference results.

– Even though application of Bayes' theorem is very often permissible from a mathematical point of view, it makes sense in practice to question, before it is implemented, the merit and the benefit of combining sources of information that may not be consistent (compatible or reconcilable) with one another. Bousquet [BOU 08] tackles this important subject and proposes a criterion quantifying the concordance between *prior* information and OF data, so as to alert the analyst if there is a conflict. We emphasize once again here the fact that, as for the frequentist framework, the Bayesian approach relies on OF data, the quality of which will consequently condition the quality of the result of the reliability study. When the number of failures is low, it is therefore only recommended to proceed to sensitivity analyses to study the robustness of the results obtained.

– The Bayesian approach provides as output the *posterior* distribution of the $\boldsymbol{\theta}$ parameters. This distribution can serve to build (Bayesian) point estimators of $\boldsymbol{\theta}$, considering for example the mode, the expectation, the median or the quantiles of the *posterior* distribution. Credibility intervals can also be constructed, which are the Bayesian counterparts of frequentist confidence intervals (see section 5.5), even though their interpretation is very different. More formally, defining a Bayesian estimator requires choosing at the outset a *posterior* risk function (and *de facto* a loss function) from which the Bayesian estimator naturally follows (see [ROB 06]).

– We specify that the Bayesian approach is much used for updating data from reliability data handbooks (see for example [EIR 00]), the content of which is vital to probabilistic safety assessments (see Chapter 3) or for study in the design stage of the dependability of a new installation. It also makes it possible to individualize the reliability performances of one device among others. For broader illustrations of use of the Bayesian approach for the main models used in the domain of equipment reliability, we advise you to read [HAM 08].

5.7. Model validation and selection

Even if the modeler has first taken quite particular care to propose (or to build) a model that he hopes will suitably characterize the family of equipment studied and the OF data available (see sections 5.2 and 5.4), then to estimate the parameters of this model using all the information accessible and by assessing the associated uncertainties (see sections 5.5 and 5.6), they should ensure retrospectively that the developed model is really faithful to the knowledge that has been used to establish it. Indeed, a model can rely on hypotheses (constant failure rate function with time, multiplicative effect of influencing factors on the failure rate function, choice of a particular family of distributions or of some explanatory variables over others, etc.) the merits of which or, at minimum, the absence of heavy impact on the final result must be verified (even though these hypotheses would rely on the results of the preliminary phase of data exploration, see section 5.3). This validation is even more important when several candidate models are available and it is necessary (often for practical reasons) to select only one, ideally "the best" of all those in competition. This problem is known by the name of model validation and selection and several categories of tools enable it to be addressed [BUR 02].

An initial possibility gathers the graphical approaches, which rely on a visual representation that helps to assess the relevance of a hypothesis or the concordance between the observed distribution of the data (without a model assumption) and the "theoretical" distribution supposed to have generated the data (generally approximated by the model $\mathcal{M}(\widehat{\boldsymbol{\theta}})$ the parameters of which have been estimated). Among them, two are very widely used for (potentially censored) OF data taken from the observation of a family of non-repairable components [LAW 03]:

– the graph overlaying on the one hand the non-parametric Kaplan–Meier estimation of the survival function $\widehat{S}_{T,KM}(t)$ (that relies solely on the OF data, see section 5.3), and on the other hand the survival function of the estimated model $S_{T,\mathcal{M}}(t|\widehat{\boldsymbol{\theta}})$ (see an example of this type of graph in Figure 5.3);

– the Weibull-plot (see one example in Figure 5.8), which relies on a linearization of the survival function, via the following relationship specific to the Weibull distribution $\mathcal{W}(\eta,\beta)$: $\ln\left(-\ln\left(S_{T,\mathcal{W}}(t|\eta,\beta)\right)\right) = \beta\ln(t) - \beta\ln(\eta)$.

Even though these graphical approaches are generally simple to implement and often useful, they present a qualitative character (since they are linked to a visual assessment), which often leads more formal approaches to be preferred.

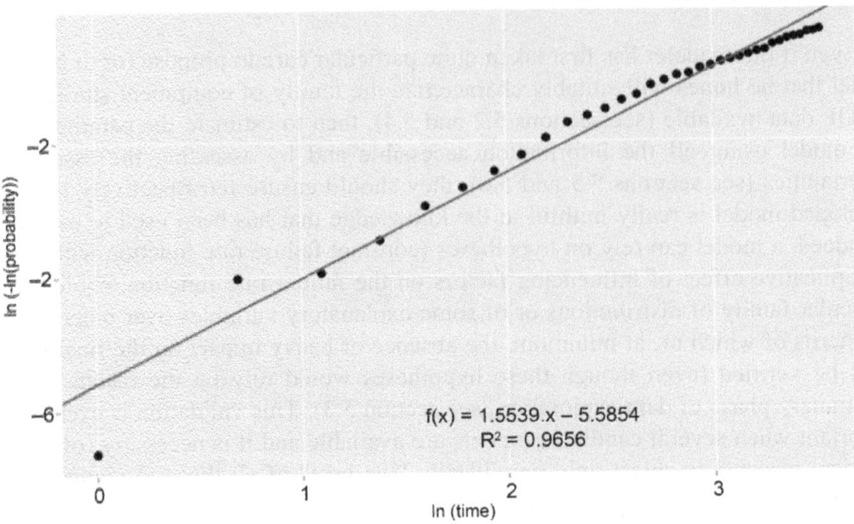

Figure 5.8. *Example of a Weibull-plot constructed on a family of non-repairable components (nuclear heat exchangers, which have the function of making it possible to collect fluid samples). The fact that the scatter plot shows an overall linear drift (despite two points apart on the left of the graph) argues in favor of the Weibull distribution as a possible probability distribution for modeling the data. This graph was generated with the 'survival' package from R software environment. For a color version of this figure, see www.iste.co.uk/lannoy/nuclear.zip*

The second category of tools covers statistical hypothesis tests. A hypothesis test is a procedure that aims to provide a decision-making rule making it possible, on the available dataset, to choose between two statistical assumptions, one written as (H.0) and called a null hypothesis (e.g. "the expected value of the random variable underlying the data is zero" or "the random variable underlying the data follows a Weibull distribution") and the other written as (H.1) and called an alternative hypothesis (e.g. the "expected value of the random variable underlying the data is equal to a particular non-zero value" or "the random variable underlying the data does not follow a Weibull distribution"). Depending on the target objective (to decide whether a statistical quantity is equal to a particular value, or on the influence of one random variable observed on the probability distribution of another, etc.), there is a very large variety of hypothesis tests (see, for example, [BAG 11a] and [BAG 11b]). We will be content in this chapter to present goodness-of-fit tests only briefly. Their aim is to help decide whether the OF data observed (e.g. operating durations collected on a family of non-repairable components, or repeated degradation increments measured on a passive component) follow a given probability distribution family (e.g. a Weibull distribution, or a gamma distribution, the parameters of which can be estimated), thus making it possible to validate how well suited the chosen hypothesized model $\mathcal{M}(\widehat{\boldsymbol{\theta}})$ is for characterizing data. More formally, if we denote the 'theoretical' distribution $\mathfrak{D}^*$ the good agreement of which with the data we wish to verify, a goodness-of-fit test means making the null hypothesis (H.0): "the data observed follow distribution $\mathfrak{D}^*$" and as an alternative hypothesis (H.1): "the data observed do not follow distribution $\mathfrak{D}^*$". To decide between (H.0) and (H.1), goodness-of-fit tests generally rely on a distance between two versions calculated from a single and same quantity (e.g. the distribution function): the version of the quantity called the "empirical version", calculated solely from the data without making any hypothesis on their distribution, and the version called the "theoretical version" of the quantity, calculated under the hypothesis that the data follow the tested distribution $\mathfrak{D}^*$. If the deviation between these two versions of the quantity is slight, this leads to the conclusion that there is some "proximity" between the data and the $\mathfrak{D}^*$ model: there are good reasons to think that the data are not "incompatible" with hypothesis (H.0) and not to reject the model's fit to the data. Conversely, if the distance is great, then we may doubt that the data are well adjusted by the model considered. For more on this theme, we will find in [KRI 19] a broad review and a comparison of the main goodness-of-fit tests for Weibull distribution (see section 5.4.2), in [BRA 01] goodness-of-fit tests dedicated to discrete models, in [BAL 13] tests adapted to Cox and AFT models to take account of influencing factors (see section 5.4.3) and in [CHA 16] (respectively, in [KAH 16]) tools to check the fit of some stochastic processes to data taken from the observation of repairable equipment (see section 5.4.4) (respectively, degradation measurements on passive components, see section 5.4.5).

The third category of tools is made up of quantitative statistical criteria for comparing several competing models and selecting the one that best fits the data available. The most well-known, such as the Akaike Information Criterion (AIC) and the Bayesian Information Criterion (BIC), are based on information theory [BUR 02]. For a given model $\mathcal{M}(\boldsymbol{\theta})$, the AIC is expressed as $\text{AIC}(\mathcal{M}(\boldsymbol{\theta})) = -2\ln\left(\mathcal{L}_{\mathcal{M}}(\text{data}|\widehat{\boldsymbol{\theta}})\right) + 2\text{d}$, with $\widehat{\boldsymbol{\theta}}$ the estimation using the maximum likelihood method of parameter $\boldsymbol{\theta}$ of the model, $\mathcal{L}_{\mathcal{M}}(\text{data}|\widehat{\boldsymbol{\theta}})$ the maximum value of the likelihood function associated with the observed data and with the model $\mathcal{M}(\boldsymbol{\theta})$ and d the dimension of $\boldsymbol{\theta}$, that is, the number of the model's parameters (see section 5.4.6). The AIC is based on the concept of information entropy, offering a relative measurement of the information lost when a given model is used to describe reality. The best model among those in competition is the one that minimizes the value of the AIC. This one introduces a penalty term via the number of parameters d of the model, which compensates the fact that the log-likelihood function is increasing according to d (which may be interpreted as a number of degrees of freedom: the higher the number of parameters in a model, the more flexible the model and the better it can fit the data). Thus, through its construction, the best model based on the AIC is a parsimonious model, which presents a "good" compromise between "likelihood" and "modeling complexity"[6]. The BIC, for its part, penalizes the number of parameters more strictly than the AIC, considering the amount n of data available. It is given by the following formula: $\text{BIC}(\mathcal{M}(\boldsymbol{\theta})) = -2\ln\left(\mathcal{L}_{\mathcal{M}}(\text{data}|\widehat{\boldsymbol{\theta}})\right) + d\ln(n)$. There again, the best model among those in competition is the one that minimizes the BIC.

Let us finish by specifying that the techniques presented in this section rely for the most part on the frequentist paradigm. However, the Bayesian approach (see section 5.6), through specific concepts such as *posterior* distribution of the parameters $\boldsymbol{\theta}$, proposes specific tools for validating and selecting models, such as the Bayes' factor (which is the Bayesian counterpart of the frequentist likelihood ratio), the Deviance Information Criterion (which can be seen as a generalization of the AIC) or the Bayesian model averaging (which consists broadly of aggregating several models by weighting them by their *posterior* probability). It must finally be specified that, aside from formal mathematical tools such as those that we have just introduced, field considerations can also help with validating or selecting models. For example, a specialist in the equipment studied can support the statistician modeler by specifying to him/her the physical factors that an industrial operator may modify (they are called maneuverable or operable factors, over which the operator has control), which can lead to a model that may be less "efficient" from a statistical point of view but easier to use and so more useful in practice, being preferred.

5.8. Case study for illustration

We present here, briefly, a real case study taken from the field of nuclear production of electricity to illustrate the concrete applicability, effective use and industrial benefit of the models and methods presented in the previous sections of this chapter. This case study is detailed in [GIL 10].

Different types of components of the primary circuit of EDF nuclear power plants are manufactured from the same alloy, subject to degradation mechanisms, the SC of which may lead to cracking. An inspection program for one of these families of (non-repairable) components made of this alloy had been defined on the basis of a physical deterministic model making it possible to assess the minimum operating time for a macroscopic crack to appear. This program relied on the list of components considered to be "precursors" (i.e. more sensitive) to the risk of cracking, completed by another list of components chosen "randomly" from among those remaining of the family studied (judged at the outset to be less subject to SC).

6 Adding additional parameters to a model can do it a disservice, by increasing the risk of "over-learning" or "over-fitting", that is, of poor generalization capacity.

The discovery of a crack during inspection of a component from this second list pushed the operator to enquire into this problem, more from a need to develop understanding of the physical phenomenon than from major safety imperatives.

To this effect, a statistical reliability analysis was carried out, independently of mechanical studies and expertise in metallurgy on the components removed from service. This study did not aim to explain the physics of the degradation phenomenon, but to provide objective elements, based on the data observed, making it possible to identify without prior field knowledge, the statistically influential physical factors and if applicable to help predict the health of components depending on the factors whose effect on the degradation by SC would be revealed to be significant from the point of view of OF. The statistical highlighting of factors relevant to the phenomenon has also been able, to some extent, to support the construction of new physical models and to strengthen understanding of the phenomenon.

The OF data available was characterized as follows. Nearly 65% of components from the family involved in the study had been the subject of non-destructive examinations, making it possible to detect the presence or absence of cracking. The presence of cracking by SC was revealed in 14% of them. For each of the components inspected, the operating duration at which the inspection was carried out was known, thus making it possible for components presenting cracking (respectively, the components not degraded) to know an upper bound (respectively, lower) of the exact delay before initiation of the cracking (left-censored data, respectively, right-censored, of the operating time before initiation). Additionally, for each component examined, the values of a number of physical factors were available, grouped into three main categories: those relating to manufacture of the component, such as the steel manufacturer, the sensitivity class of the material, the chemical composition of the cast, the sheet metal from which it was made, the temperature and duration of heat treatment, the type of chamfer, the strain hardening, the thickness of the component, etc.; those relating to mechanical effects, such as the yield strength or the effect of hydraulic tests; those relating to the component's usage profile, such as the operating duration and the temperature of the operating environment.

Analysis aiming to identify factors statistically influential on the cracking phenomenon was carried out in three stages. First, a descriptive statistical study of the OF data was carried out, mainly via statistical tests (see section 5.7) making it possible to identify a potential statistical link between the variable of interest characterizing the presence or absence of cracking on the components and each of the covariates considered separately (search for statistical correlations). This analysis made it possible to become familiar with the OF and the factors available and to detect the first statistical trends on the physical parameters that could be involved in the cracking process. Second, the knowledge available from OF was modeled relying on AFT models with covariates (see section 5.4.3), aiming to describe the distribution of the initiation time of cracking through SC. Of all the models constructed, the “best” from a statistical point of view was selected by minimizing the Akaike Information Criterion (see section 5.7). This model, a Weibull type model, is represented in Figure 5.9 through its estimated cumulative distribution function, plotted for three different values (one penalizing value, one reference value and one favorable value, all things being equal, moreover) of one of the physical factors identified as influencing the initiation time of cracking and considered in the accelerated life model. It is easy to observe the importance of this covariate on the degradation initiation delay.

In a third and final stage, the model selected above was used with a view to predicting the operating duration before cracking. Thus, the study endeavored to assess two indicators: on the one hand, the average time before initiation of the degradation, on the other hand, the average remaining useful life, that is, the average operating duration before initiation of the degradation given that the inspection carried out on the component at a given time concluded that cracking was absent. This last indicator served to establish an inspection program taking account of the individual risk of cracking of each component depending on its own characteristics.

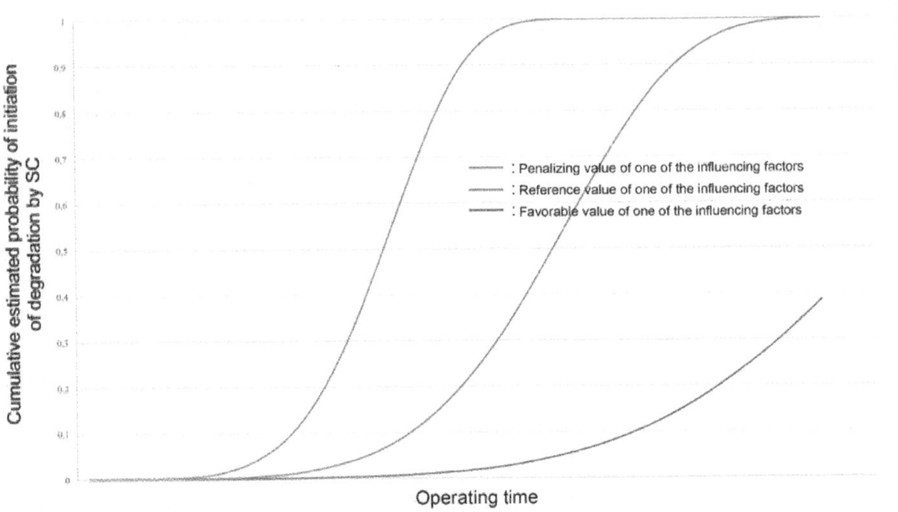

Figure 5.9. *Plot of the estimated cumulated distribution function of the initiation time of degradation by SC for three different values of one of the physical factors identified as influential. For a color version of this figure, see www.iste.co.uk/lannoy/nuclear.zip*

5.9. Openings and prospects for R&D

Several openings and prospects for R&D have already been mentioned in the previous sections of this chapter. Also, and in "connection" with Chapter 7, we will limit ourselves here to mentioning big data as a support for assessing the reliability of equipment in operation. Indeed, because of technological development (less costly and more reliable instrumentation of equipment, generalization of connected objects, miniaturization and increase of the performances of means of controlling the equipment and collecting information, increased data storage capacities, etc.), big data will progressively become a reality for the industrial world and it offers promising openings, both in terms of methods and applications. Some previous studies [MEE 14, HON 18, IMD 20] discuss these opportunities and the associated challenges, which consist mainly of integration, into classical reliability models (such as those presented in this chapter), new data (so far inaccessible), that is simultaneously streaming, multi-dimensional, voluminous and very varied in type and structure (time series of measurements of many usage and environment parameters taken from a network of sensors, images or films from visual inspections by drone, acoustic signals recorded by sensors, written maintenance reports, etc.). Learning to make use of this new information will make it possible to diagnose more accurately and in a more individualized way the health of equipment, to follow more precisely the evolution of operating, environment and maintenance conditions of devices and to assess more rigorously their effect on reliability, predict more accurately the future behavior of the equipment (prognosis and health management [PHM]) and consequently to adapt maintenance, thus improving the safety and performance of industrial installations overall.

Some previous studies [HON 13, HON 15] perfectly illustrate this trend, by adapting standard models (as it happens, an accelerated life model and a degradation path model, introduced in sections 5.4.3 and 5.4.5) to take account of time-dependent covariates (for which predictive evolution models are also constructed, so as to be able to carry out prognosis on the health of equipment). Lee *et al.* [LEE 18] (among other recent works) consider the logic of big data even further and propose an approach to reliability analysis that is conceptually much distant from the one presented here (in which probabilistic modeling takes a central role). Indeed, the authors do not consider a parametric stochastic model such as those described in this chapter (which rely for the most part on modeling hypotheses and precise expressions of the mathematical functions characterizing the probability distribution of the operating duration until failure), but build a deep neural network to infer (learn) the lifetime of the components studied. Such machine learning algorithms require a large amount of data to be effective, and they do not easily make it possible to quantify the precision of the results obtained as an output and sometimes they lack interpretability ("black box" effect). However, the authors highlight their capacity, which is moreover undeniable, to integrate nonlinear behaviors and to capture information contained in the data that would be impossible, or difficult, to make explicit in a classic parametric probabilistic model.

This underlying trend, which aims to use ever more voluminous and heterogeneous data, militates for progress on the theme of information quality, which is absolutely essential, as we have been at pains to repeat throughout this chapter, but which, in the end, is the subject of little academic research and would certainly merit more attention (original work will be cited to this effect [KEN 14]).

5.10. Software tools

It is difficult to draw up a list of existing software tools to help carry out a statistical study on the reliability of industrial equipment in operation, and even more so to keep it up to date. We will therefore content ourselves with indicating that programming languages under free license Python (https://www.python.org/) and R (https://www.r-project.org/, which was used to produce the figures in this chapter) are currently the two main software references in the domain of statistics, with a community of contributors/developers, which make available to users a broad variety of libraries (also called packages) dedicated to disciplinary or cross-disciplinary domains of study (econometrics, time series, extreme values, reliability, hydrology, social sciences, etc.). The two online resources, such as https://pypi.org/ (for Python) and https://cran.r-project.org/web/views/ (for R), make it possible to guide users to libraries on their subject of interest[7]. In addition, the web page https://en.wikipedia.org/wiki/List_of_statistical_software provides a wide list of statistical software tools (commercial and free), of which only a part relates to the field of equipment reliability.

5.11. References

[ALM 17] AL MASRY Z., MERCIER S., VERDIER G., "A condition-based dynamic maintenance policy for an extended gamma process", *10th International Conference on Mathematical Methods in Reliability*, Grenoble, 2017.

[ASS 18] ASSOCIATION FRANÇAISE DE NORMALISATION, "Maintenance – Maintenance terminology", Association Française de NORmalisation, NF X60-319, NF EN 13306:2018, 2018.

7 For the domain of equipment reliability more specifically, we can cite, for example, the following libraries: Python "lifelines", "PySurvival", "python-asurv", "scikit-survival", "PyMC3" and "ArviZ" and R packages "survival", "EWGoF", "NHPoisson", "VAM", "yuima" and "rstanarm".

[BAC 98] BACHA M., CELEUX G., IDEE E. *et al.*, *Estimation de modèles de durées de vie fortement censurées*, Eyrolles, Paris, 1998.

[BAG 01] BAGDONAVICIUS V., NIKULIN M.S., *Accelerated Life Models – Modeling and Statistical Analysis*, Chapman and Hall/CRC, Boca Raton, FL, 2001.

[BAG 11a] BAGDONAVICIUS V., KRUOPIS J., NIKULIN M.S., *Nonparametric Tests for Complete Data*, ISTE, London, and Wiley, Hoboken, 2011.

[BAG 11b] BAGDONAVICIUS V., KRUOPIS J., NIKULIN M.S., *Non-parametric Tests for Censored Data*, ISTE, London, and Wiley, Hoboken, 2011.

[BAL 13] BALAKRISHNAN N., CHIMITOVA E., GALANOVA N. *et al.*, "Testing goodness of fit of parametric AFT and PH models with residuals", *Communications in Statistics – Simulation and Computation*, vol. 42, no. 6, pp. 1352–1367, 2013.

[BAR 65] BARLOW R.E., PROSCHAN F., *Mathematical Theory of Reliability*, John Wiley & Sons, Hoboken, 1965.

[BAY 63] BAYES T., "An essay towards solving a problem in the doctrine of chances", *The Philosophical Transactions of the Royal Society*, vol. 53, pp. 370–418, 1763.

[BER 01] BERTHOLON H., Une modélisation du vieillissement, PhD Thesis, Université Joseph Fourier, Grenoble, 2001.

[BOR 16] BORDES L., MERCIER S., REMY E. *et al.*, "Partially observed competing degradation processes: modeling and inference", *Applied Stochastic Models in Business and Industry*, vol. 32, no. 5, pp. 677–696, 2016.

[BOU 08] BOUSQUET N., "Diagnostics of prior-data agreement in applied Bayesian analysis", *Journal of Applied Statistics*, vol. 35, no. 9, pp. 1011–1029, 2008.

[BRA 01] BRACQUEMONT C., Modélisation stochastique du vieillissement en temps discret, PhD Thesis, INPG, Grenoble, 2001.

[BUR 02] BURNHAM K.P., ANDERSON D.R., *Model Selection and Multimodel Inference – A Practical Information-theoretic Approach*, 2nd ed., Springer, New York, 2002.

[CHA 16] CHAUVEL C., DAUXOIS J.Y., DOYEN L. *et al.* "Parametric bootstrap goodness-of-fit tests for imperfect maintenance models", *IEEE Transactions on Reliability*, vol. 65, no. 3, pp. 1343–1359, 2016.

[COO 07] COOK R.J., LAWLESS J.F., *The Statistical Analysis of Recurrent Events*, Springer, New York, 2007.

[COR 12] CORSET F., DOYEN L., GAUDOIN O., "Bayesian analysis of ARA imperfect repair models", *Communications in Statistics – Theory and Methods*, vol. 41, pp. 3915–3941, 2012.

[COX 72] COX D.R., "Regression models and life-tables", *Journal of the Royal Statistical Society. Series B (Methodological)*, vol. 34, no. 2, pp. 187–220, 1972.

[CRO 12] CROWDER M.J., *Multivariate Survival Analysis and Competing Risks*, Chapman and Hall/CRC, New York, 2012.

[DIA 2018] DIAS L.C., MORTON A., QUIGLEY J., *Elicitation – The Science and Art of Structuring Judgement*, Springer International Publishing, Cham, 2018.

[DOY 04] DOYEN L., GAUDOIN O., "Classes of imperfect repair models based on reduction of failure intensity or virtual age", *Reliability Engineering and System Safety*, vol. 84, no. 1, pp. 45–56, 2004.

[EFR 81] EFRON B., "Censored data and the bootstrap", *Journal of the American Statistical Association*, vol. 76, no. 374, pp. 312–319, 1981.

[EFR 94] EFRON B., TIBSHIRANI R.J., *An Introduction to the Bootstrap*, Chapman and Hall/CRC, Boca Raton, 1994.

[EGG 23] EGGENBERGER F., POLYA G., "Uber die Statistik verketter Vorgänge", *Zeitschrift für angewandte Mathematik und Mechanik*, vol. 3, pp. 279–289, 1923.

[EIR 00] EIREDA, *European Industry Reliability Data Bank*, Crete University Press, Heraklion, 2000.

[GAM 11] GAMIZ M.L., KULASEKERA K.B., LIMNIOS N. *et al.*, *Applied Nonparametric Statistics in Reliability*, Springer, London, 2011.

[GER 17] GERVILLE-REACHE L., COUALLIER V., BAYLE F., Approches comparées sur l'estimation d'une probabilité de défaillance : cas des échantillons totalement censurés, available at: https://hal.archives-ouvertes.fr/hal-01448201, 2017.

[GIL 10] GILLES E., LEMAIRE E., DEFORGE D., *Apport de la statistique pour la compréhension et la prévision de phénomène de dégradation*, Congrès λμ17, La Rochelle, 2010.

[HAM 08] HAMADA M.S., WILSON A., REESE C.S. *et al.*, *Bayesian Reliability*, Springer, New York, 2008.

[HON 13] HONG Y., MEEKER W.Q., "Field-failure predictions based on failure-time data with dynamic covariate information", *Technometrics*, vol. 55, no. 2, pp. 135–149, 2013.

[HON 15] HONG Y., DUAN Y., MEEKER W.Q. *et al.*, "Statistical methods for degradation data with dynamic covariates information and an application to outdoor weathering data", *Technometrics*, vol. 57, no. 2, pp. 180–193, 2015.

[HON 18] HONG Y., ZHANG M., MEEKER W.Q., "Big data and reliability applications: the complexity dimension", *Journal of Quality Technology*, vol. 50, no. 2, pp. 135–149, 2018.

[IMD 20] IMDR, *Projet IMdR P17-4. Big Data in Reliability, T3 – Impact du Big Data sur l'estimation de la fiabilité d'un composant, d'une fonction ou d'un système*, Institut pour la Maîtrise des Risques, Cachan, 2020.

[INT 97] INTERNATIONAL ELECTROTECHNICAL COMMISSION, *Reliability Data Analysis Techniques. Procedures for Comparison of Two Constant Failure Rates and Two Constant Failure (Event) Intensities*, edition 1.0, International Electrotechnical Commission, 61650:1997, Geneva, 1997.

[INT 01a] INTERNATIONAL ELECTROTECHNICAL COMMISSION, *Equipment Reliability Testing. Part 4: Statistical Procedures for Exponential Distribution – Point Estimates, Confidence Intervals, Prediction Intervals and Tolerance Intervals*, International Electrotechnical Commission, 60605–4:2001, 2001.

[INT 01b] INTERNATIONAL ELECTROTECHNICAL COMMISSION, *Dependability Management. Part 3-5: Application Guide – Reliability Test Conditions and Statistical Test Principles*, edition 1.0, International Electrotechnical Commission, 60300-3-5:2001, 2001.

[INT 04] INTERNATIONAL ELECTROTECHNICAL COMMISSION, *Reliability Growth. Statistical Test and Estimation Methods*, edition 2.0, International Electrotechnical Commission, 61164:2004, 2004.

[INT 06a] INTERNATIONAL ELECTROTECHNICAL COMMISSION, *Equipment Reliability. Reliability Assessment Methods*, edition 1.0, International Electrotechnical Commission, 62308:2006, 2006.

[INT 06b] INTERNATIONAL ELECTROTECHNICAL COMMISSION, *Application of Markov Techniques*, edition 2.0, International Electrotechnical Commission, 61165:2006, 2006.

[INT 07] INTERNATIONAL ELECTROTECHNICAL COMMISSION, *Equipment Reliability Testing. Part 6: Tests for the Validity and Estimation of the Constant Failure Rate and Constant Failure Intensity*, edition 3.0, International Electrotechnical Commission, 60605-6:2007, 2007.

[INT 08] INTERNATIONAL ELECTROTECHNICAL COMMISSION, *Weibull Analysis*, edition 2.0, International Electrotechnical Commission, 61649:2008, 2008.

[INT 10] INTERNATIONAL ELECTROTECHNICAL COMMISSION, *Functional Safety of Electrical/electronic/programmable Electronic Safety-related systems*, edition 2.0, International Electrotechnical Commission, 61508:2010, 2010.

[INT 13] INTERNATIONAL ELECTROTECHNICAL COMMISSION, *Power Law Model. Goodness-of-fit Tests and Estimation Methods*, edition 2.0, International Electrotechnical Commission, 61710:2013 2013.

[INT 16] INTERNATIONAL ELECTROTECHNICAL COMMISSION, *Mathematical Expressions for Reliability, Availability, Maintainability and Maintenance Support Terms*, edition 2.0, International Electrotechnical Commission, 61703:2016, 2016.

[KAH 16] KAHLE W., MERCIER S., PAROISSIN C., *Degradation Processes in Reliability*, ISTE Ltd, London and John Wiley & Sons, Hoboken, 2016.

[KAL 11] KALBFLEISCH J.D., PRENTICE R.L., *The Statistical Analysis of Failure Time Data*, John Wiley & Sons, Hoboken, 2011.

[KAP 58] KAPLAN E.L., MEIER P., "Non parametric estimation from incomplete observations", *Journal of the American Statistical Association*, vol. 53, pp. 457–481, 1958.

[KEN 14] KENETT R.S., SHMUELI G., "On information quality", *Journal of the Royal Statistical Society, Series A*, vol. 177, no. 1, pp. 3–38, 2014.

[KIJ 89] KIJIMA M., "Some results for repairable systems with general repair", *Journal of Applied Probability*, vol. 26, pp. 89–102, 1989.

[KRI 19] KRIT M., GAUDOIN O., REMY E., "Goodness–of–fit tests for the Weibull and extreme value distributions: a review and comparative study", *Communications in Statistics - Simulation and Computation*, pp. 1888–1911, 2019.

[LAW 03] LAWLESS J.F., *Statistical Models and Methods for Lifetime Data*, 2nd ed., Wiley, John Wiley and Sons, Hoboken, 2003.

[LEE 18] LEE C., ZAME W.R., YOON J. *et al.*, "Deephit: a deep learning approach to survival analysis with competing risks", *Thirty-Second AAAI Conference on Artificial Intelligence*, New Orleans, February 2–7, 2018.

[MEE 98] MEEKER W.Q., ESCOBAR L.A., *Statistical Methods for Reliability Data*, Wiley, John Wiley and Sons, Hoboken, 1998.

[MEE 14] MEEKER W.Q., HONG Y., "Reliability meets big data: opportunities and challenges", *Quality Engineering*, vol. 26, pp. 102–116, 2014.

[MEE 17] MEEKER W.Q., HAHN G.J., ESCOBAR L.A., *Statistical Intervals: A Guide for Practitioners and Researchers*, 2nd ed., Wiley, Hoboken, 2017.

[PAR 14a] PAROISSIN C., SALAMI A., "Failure time of non homogeneous gamma process", *Communications in Statistics–Theory and Methods*, vol. 43, no. 15, pp. 3148–3161, 2014.

[PAR 14b] PAROISSIN C., REMY E., VERRIER V., "Inférence statistique pour un modèle Markovien de dégradation avec covariables dépendantes du temps", *Journal de la Société Française de Statistique*, vol. 155, no. 1, pp. 99–116, 2014.

[PAS 15] PASANISI A., ROERO C., BOUSQUET N. *et al.*, "On the practical interest of discrete Inverse Pólya and Weibull-1 models in industrial reliability studies", *Quality and Reliability Engineering International*, vol. 31, pp. 1161–1175, 2015.

[PEN 15] PENG C.Y., "Inverse Gaussian processes with random effects and explanatory variables for degradation data", *Technometrics*, vol. 57, no. 1, pp. 100–111, 2015.

[PRO 11] PROCACCIA H., FERTON E., PROCACCIA M., *Fiabilité et maintenance des matériels industriels réparables et non réparables*, Tec et Doc – Lavoisier, Paris, 2011.

[RAM 19] RAM M., *Reliability Engineering: Methods and Applications*, CRC Press, Boca Raton, 2019.

[REM 13] REMY E., CORSET F., DESPREAUX S. *et al.*, "An example of integrated approach for the technical and economic optimization of maintenance", *Reliability Engineering and System Safety*, vol. 116, pp. 8–19, 2013.

[RIN 08] RINNE H., *The Weibull Distribution: A Handbook*, CRC Press, Boca Raton, 2008.

[ROB 06] ROBERT C.P., *Le choix bayésien : principes et pratique*, Springer, Paris, 2006.

[THE 00] THERNEAU T.M., GRAMBSCH P.M., *Modeling Survival Data: Extending the Cox Model*, Springer, New York, 2000.

[VAN 09] VAN NOORTWIJK J.M., "A survey of the application of gamma processes in maintenance", *Reliability Engineering and System Safety*, vol. 94, pp. 2–21, 2009.

[WAS 68] WASAN M., "On an inverse Gaussian process", *Scandinavian Actuarial Journal*, pp. 69–96, 1968.

[WIE 10] WIENKE A., *Frailty Models in Survival Analysis*, Biostatistics Series, Chapman & Hall/CRC, Boca Raton, 2010.

[WOO 19] WOO S., *Reliability Design of Mechanical Systems – A Guide for Mechanical and Civil Engineers*, 2nd ed., Springer, Singapore, 2019.

[YE 14] YE Z.S., CHEN N., "The inverse Gaussian process as a degradation model", *Technometrics*, vol. 56, no. 3, pp. 302–311, 2014.

[YE 15] YE Z.S., XIE M., "Stochastic modelling and analysis of degradation for highly reliable products", *Applied Stochastic Models in Business and Industry*, vol. 31, no. 1, pp. 16–32, 2015.

6

The Human and Organizational Dimensions of Reliability and Nuclear Safety

Nuclear accidents have drawn attention to consideration of human and organizational dimensions to improve nuclear safety. Theories on the occurrence of accidents and on human and organizational reliability have emerged since the 1980s. These theories as well as approaches and methods taken from ergonomics, psychology and sociology have allowed the development of practices relating to human and organizational factors (HOF) that have been consolidated and spread. Hundreds of expert and research studies have thus been conducted by numerous specialists in France. Some key points relating to human or organizational factors, or more cross-cutting aspects, from a historical or theoretical point of view or in terms of the state of the art of practices are developed in this chapter: they illustrate some applications and methodological principles, and open up perspectives for improvement in the domain of risk management.

6.1. Introduction and historical context in the nuclear field

Human and organizational dimensions of nuclear reliability and safety have a long history. Stakeholders in the French nuclear industry (designers, operators, sub-contractors, experts, safety authorities and civil society) thus have more than 40 years of research and practical work in design, operation, maintenance, expertise and management. Good practices, but also failures and controversies, have supplied the development of approaches and methods for managing risk.

Chapter written by Nicolas DECHY, Yves DIEN and Jean-François VAUTIER.

The nuclear industry was designed and begun without expertise and without in-depth research on the human and organizational dimensions, before the disciplinary field of reliability was even formed. If worries, or indeed research, on the prevention of human errors or on adequate organization for managing the risks of nuclear reactors led to concrete measures from the beginnings of the industry in France, it was above all the result of accidents (Three Mile Island in 1979, Chernobyl in 1986 and Fukushima in 2011) which would put these dimensions on center stage [BER 14, COU 20]. These accidents were in fact real shocks for the global nuclear industry. They put the use of nuclear energy into question in some countries. They triggered, in France and in other countries, significant efforts that took concrete form in studies, research and the refinement of practical arrangements for design and operation with the aim of limiting the risks of human error and organizational failure. Until the 1970s, and as in other industries [WIL 98], risk management relied essentially on a reliable technical design. The accidents, to some extent, called into question the representations that formed the basis of risk management modeling. Analytical fields widened progressively (schematically from one decade to another) from human factors (HF) to HOF and cultural factors, that is, from risk management to risk governance (taking account of relationships with safety authorities, sub-contractors, French and foreign operators and stakeholders).

The international context was therefore decisive for consideration of these dimensions by the French nuclear industry. This evolution was not uniquely observed in the nuclear domain, it happened in other sectors also after the occurrence of serious accidents (see, for example, the consequences, for aviation, of the collision of two Boeing 747 in March 1977 on the island of Tenerife in the Canaries). International cooperation resulting from the extent of the consequences of nuclear accidents (*an accident somewhere is an accident everywhere*, attributed by the director of the IAEA[1], Mohamed El Baradei [ELB 05] to his predecessor Hans Blix) was emphasized, after nuclear accidents, with many collective works involving operators and safety authorities, especially within the IAEA and the OECD[2] since the 1980s. Similarly, operators developed the sharing of good practice within WANO[3], or pooled their study and research resources in the United States (INPO[4]). Western European (including Central European countries and Scandinavian countries) safety authorities were also grouped together within WENRA[5], as well as their technical supports (i.e. institutes of expertise) within ETSON[6]. Each of these bodies produced reports, guides for better consideration of HOF in risk

1 International Atomic Energy Agency.

2 Organisation for Economic Cooperation and Development.

3 World Association of Nuclear Operators.

4 Institute of Nuclear Power Operations.

5 Western European Nuclear Regulators Association.

6 European network of Technical Support Organization.

management, to which must be added the establishment of standards and many scientific and technical publications as well as the organization of scientific conferences [IAE 16b].

In France, since the start of the 1980s, teams of specialists were formed by EDF[7] (at the Nuclear Generation Division and at the Research and Development Laboratories), *Orano* (formerly *Areva, Cogema*), *Framatome, Edvance* and the IRSN[8] (formerly IPSN), the technical support of the French Regulator. HOF specialists were recruited as well by the Alternative Energies and Atomic Energy Commission (CEA) and the Nuclear Safety Authority (ASN). Today, the disciplines in Human and Social Sciences (HSS) are mainly ergonomics, sociology and psychology but also address other fields such as anthropology, management and political sciences. It leads that in 2020, France had around 100 HOF/HSS experts working full time in the domain of nuclear safety: researchers at EDF R&D, experts in the central departments at EDF, a "human factors consultant" for every pair of nuclear reactors, etc. Experts at IRSN [ROL 09], safety authorities, Edvance and Orano can also be counted. In addition, some specific studies are carried out by academics or consultants. In addition, hundreds of relay-actors are trained in HOF practices (e.g. reliability practices, event root cause analysis, and so on). HOF networks [VAU 11] were put in place to share HOF practices, especially from the 2000s by operators and more recently by safety authorities. A growing number of managers, engineers and nuclear technicians are becoming sensitized to the human and organizational dimensions of risk management.

6.2. Definition of the human and organizational dimensions of dependability and nuclear safety

In France, the expression "Human and Organizational Factors" (HOF) is a notion defined in article 1.3 of the decree of February 7, 2012 setting general rules relating to basic nuclear installations. HOFs are described there as "factors that have an influence on human performance, such as skills, the working environment, task characteristics, and the organization".

This expression is also an eponymous discipline that forms part of HSS (human and social sciences). Its founding principles are mentioned in the definition presented. They rely on different models of the scientific and technical literature of the HSS domain (especially [LEP 01, REA 06, DAN 10, AIE 13]):

7 Électricité De France.

8 Institut de Radioprotection et de Sûreté Nucléaire (Institute for Radiation Protection and Nuclear Safety).

– Jacques Leplat's model [LEP 01] proposes explaining the occurrence of human performance as the result of the influence of working conditions known as internal working conditions (linked to the human operators involved) and external working conditions (linked to the task, to the tools used, to the environment and the organization of work, etc.) as well as interactions between these conditions.

– The series of models by James Reason (Swiss Cheese Model (SCM) [REA 06]) proposes explaining the occurrence of failures from sets of factors organized differently depending on the SCM considered. Then, the third example of the series (SCM3) presents a three-level structure ranging from inappropriate actions or human failures ("unsafe acts") downstream of the causal chain, to local workplace factors, themselves exposed upstream of this chain to organizational factors.

– The approach of the IAEA [IAE 13] defines HOFs as interactions between individuals, technology and organizations. In this framework, it is a systemic approach according to the IAEA [IAE 16a].

This definition of the decree of February 7, 2012 situates the performance of people at work in a local and global explanatory context. On the local side, human performance is therefore causally linked to a work situation that includes not only the internal characteristics of individuals ("skills" for example) but also characteristics external to the individuals involved (such as "the work environment" and "the characteristics of tasks"). On the global side, the definition indicates the "organizational" factors. We note that the working environment covers technical factors (tools, etc.) used by individuals. Finally, we remark that this definition is applied just as much to the work of a human operator operating a remote manipulator system to the work of the manager of a nuclear plant preparing a safety report. To summarize, three elements should be retained:

– There is a need, within the HOF framework to "think global". The definition in the decree focuses at once on the work situation and the organization. This means that the occurrence of a performance or human error should always be linked to an explanatory context that is both local and global. A theory of the "spontaneous generation" of human performance or human error cannot therefore be mentioned in the field of human activities.

– Consequently, focusing on human error should also lead to investigation into the explanatory elements behind this error, for which its occurrence is supported [VAU 18b]; to summarize, recall that "error is not the cause but the consequence..."; "a symptom more than a syndrome".

– The interactions between different factors within a local or global level as well as interactions between local or global factors are at the heart of the explanation behind the occurrence of human failures (see the notion of coherence or adaptation section 6.7.2). HOF specialists are interested especially in the interactions between, on the one hand, the working environment, tools, task characteristics, skills, etc., within a work situation and, on the other hand, the structure and operational rules of the organization that will make it possible in particular to develop and maintain work situations, leading at last to human and organizational performances.

We also note that the French regulator (ASN) also introduced the concept of social, human and organizational factors (SHOF), which position even more clearly the importance of the socio-organizational context, especially since the creation in 2012 of a Steering Committee on SHOF, an instance and space for exchange about SHOF for stakeholders involved in nuclear safety. Since use of the term *human factors* by British aviation during the Second World War [ADA 06], the established terminology is "human", "organizational" and sometimes "social" or "societal" "factors". We note that this vocabulary is coherent with engineering approaches that aim to mark out the conditions and factors that exercise influence on a situation or on a configuration from a causal point of view. Nevertheless, the "dimension" terminology retained in this chapter makes it possible to emphasize the need to open up fields for analysis with sociotechnical, multi-causal and consequently multidimensional (physical, human, organizational, social, etc.) perspectives without making assumptions on the nature of the causality.

6.3. Theories on accidents and reliability

It makes sense to remark that causes relating to HOFs should be considered to be complex, probabilistic, and stand out from mechanical causalities [MOR 77, LEM 77, DEC 12, VAU 18b]. HOFs are thus considered as influencing factors on complex, multi-causal situations or configurations [VAU 18a]. Analyses of accidents, incidents and crises have shown that:

> any event is initiated by immediate, direct, technical causes and/or human causes ('Error') BUT its occurrence and/or its development is induced, favored and precipitated by causes and underlying organizational conditions (complex factors) ([DIE 06], p. 148).

These studies showed that a vast majority of events can be seen as the final point of a process that degrades safety. An event is very rarely an "unexpected combination of circumstances" or an "act of God". In fact, an accident happens at the end of an "incubation period" [TUR 78] during which events, signals (weak or strong) occur and are not perceived and/or treated adequately with regard to their potential threat for safety. Any industrial system faces factors that impact safety positively and negatively. The life of an industrial system, from a safety perspective, can be seen as a continuous tension between organizational resilience factors (ORF) and pathogenic organizational factors (POF). It is possible to consider that an accident occurs when POFs dominate ORFs [DIE 06]. This definition integrates several theoretical references that are recalled in the following.

The President's Enquiry Commission on the Three Mile Island accident (TMI) ([KEM 79] p. 10) noted that *the fundamental cause of the accident was an "operator error"*. Although the concept of human error had already been mentioned from the start of the 1970s, it was on this occasion that the "concept" of human error appeared in the field of nuclear safety, a concept that would bear fruit in many other industrial sectors. Thus, the "human at work" is seen from a negative perspective, and attention comes to bear on the "front line" actors who act directly on the process. It should be noted that high-risk industries did not consider the aftermath of the Enquiry Commission report that "cleared" the operators as the only ones responsible for the event by drawing attention to organizational causes. The report underlines that:

> while the major factor that turned this incident into a serious accident was inappropriate operator action, many factors contributed to the action of the operators, such as deficiencies in their training, lack of clarity in their operating procedures, failure of organizations to learn the proper lessons from previous incidents, and deficiencies in the design of the control room. These shortcomings are attributable to the utility, to suppliers of equipment, and to the federal commission that regulates nuclear power. Therefore – whether or not operator error "explains" this particular case – given all the above deficiencies, we are convinced that an accident like Three Mile Island was eventually inevitable ([KEM 79], p. 11).

This accident occurred in parallel with the beginning of studies, mainly in English on the causes of industrial events that made it possible to develop accident occurrence models. Barry Turner [TUR 78] identified similar phenomenon in the approximately hundred of events he studied in various industrial sectors. He proposed an accident development model formed of six stages, among which: first of all a theoretically normal period of operation, which is a period of shared beliefs on the system's operation, on its risks and on the means for managing them. Rules, procedures, practices, to ensure the system's safe operation, are associated with these beliefs. For B. Turner, safety management did not therefore rely only on knowledge of the system. It is interesting to note that the accident would not be defined in technical terms (physical impact) but in sociological terms (failure of existing cultural beliefs on the subject of safety and risks). Through the system's life, there begins an "incubation period" during which we see an accumulation of a set of events that are not perceived, not raised and which are in opposition with shared beliefs. After the event has been triggered, after its management and the recovery of the situation, opens up the period of *complete cultural readjustment* during which beliefs and standards of caution are adapted so as to be compatible with the new vision/understanding of the world. This modeling of the accident draws out the dynamic of an event, which is the result of a history, and which does not happen suddenly, at random. He also shows that the incubation period is favorable to prevention if the occurrence of minor events is perceived as an early warning signal, a symptom of the degradation of the safety level.

The accident at TMI has been the subject of a collection of scientific articles [SIL 82], which include among others, two articles that gave rise to two different schools of thought in industrial safety. On the one hand, the article by Charles Perrow which argued that a serious accident was inevitable following some conditions [PER 82], and, on the other hand, an article by Todd LaPorte who specified that analysis of good practice could improve safety [LAP 82].

Perrow [PER 84] published a book that gave rise to a stream of research called "normal accident theory" [SAG 93]. For him, the accident was embedded into the system's technical dimension. The accident "is called normal because it is inherent to the characteristics of highly coupled and complex systems, and it cannot be avoided" [PER 82]. It is the result of unforeseen interaction of relatively minor failures. Two independent dimensions should be considered for the occurrence of a normal accident: the complexity of a system covers all the interactions possible between equipment and/or components and/or functions; the coupling depicts the degree of dependence and flexibility between the different elements and the different sequences of the system. Hopkins [HOP 01] would note that this model was rather theoretical and would contest it since "the normal accident" is not often observed. Regular study of industrial accidents shows that these are for the most part avoidable because of the occurrence of early warning signals but that they are still not avoided [LLO 18]. Nevertheless, after the accident at TMI, the benefit of this theory was on the one hand to oppose the simplistic view of the "operators" as the main cause of the event's occurrence and, on the other hand, it made it possible to call into question a penchant of engineers, which is to add technical systems to improve safety, while at the same time these add to the overall complexity of the system. It draws attention to systemic causes.

The article by LaPorte [LAP 82] lays the foundations of a research movement on "High Reliability Organizations" (HRO)). The movement was structured at the University of California-Berkeley mainly around Todd LaPorte, Karlene Roberts, Paula Consolini, Gene Rochlin and Karl Weick. These researchers (see, for example, [ROB 93, BOU 01, WEI 01]) found that there were many high-risk businesses while at the same time, there were few accidents. Their hypothesis, which they would test in different sectors, was to consider that these businesses were endowed with an organization that had no choice other than to promote reliable operation in view of the potentially disastrous consequences of an accident. Their research revealed that management is very much implicated in risk prevention; there is a search for agreement on the goals sought during arbitrations; organizational redundancies are created; there is a search for equilibrium between centralization and decentralization; continuous development sessions and crisis exercises are organized. Weick and Sutcliffe [WEI 01] propose five characteristics of HROs:

– preoccupation with failures and not only to successes; malfunctioning that should be treated as symptoms of degradation within the system and which could give rise to disastrous consequences;

– reluctance to simplify interpretations, so as to have a vision of the situation that is both more complete and more nuanced;

– managers sensitivity to (daily) operations so that they can develop a "situational awareness", that is, some knowledge and perception of the difficulties encountered by field operators;

– a commitment to resilience so that a disfunction does not amplify to the point of destabilizing the system totally;

– deference to expertise, that is, knowing how to delegate and rely on the individual who has the best knowledge of the situation.

James Reason, a psychologist, distinguished two types of failure: latent failures such as design or maintenance faults that only await specific conditions to be implicated in an event sequence, and active failures, such as front line operator errors, which have an immediate effect [REA 90]. These failures are integrated into an accident model [REA 90] called the "Swiss Cheese" model which had great success with risk prevention specialists. Like Turner, the historic dimension in the occurrence of an accident appears with the concept of the "accident trajectory". Risk management is ensured by a number of defense-in-depth barriers in series. But each barrier is riddled with "holes" as it has latent fault or defects. If the holes – in all the barriers – are "aligned", then the accident occurs: this is the metaphor of the model made of slices of Swiss Cheese. Reason developed a new version of his new accident model [REA 97] that distinguishes three levels (see section 6.2): dangerous acts, factors linked to work situations and organizational factors. This is coherent with sociotechnical modeling of a system by Jens Rasmussen [RAS 97] that distinguishes six levels of interaction: the industrial (reactor) process; the personnel who work and are involved in the process; management who supervise and create frameworks; the business that fixes a policy and follows indicators; control authorities who inspect, and finally the governmental level where regulations are voted for and established. This sociotechnical system is not static since it evolves in a context of internal and external constraints (changes of public opinion, economic conditions, competencies and technology). The organizational paradigm [LLO 10b] is also integrated in this modeling of the accident: these are the organizational factors that lie "at the upper levels" of the system that lead to negative factors on work situations and favor the occurrence of dangerous actions. They also have the property of acting directly on defenses (annihilation or attenuation) and of leading to dangerous acts. Although they remain unchanged, efforts toward improvement at workplace level and at the level of human activities would remain in vain. Reason [REA 97] stated the organizational factors represent the domain of ownership for improving risk management.

Following the accident of the space shuttle Challenger in 1986, the sociologist Diane Vaughan [VAU 96] led an ethnographic study, spending some months inside NASA to study organizational phenomena and studying the archives of this agency over several years. Her analysis of the accident showed that the latter was anchored in the "banality of organizational life", that it was promoted by an environment of scarcity and competition (production pressures). The researcher described how deviations within the organization were transformed into acceptable behavior, and how a collective standard was gradually built accepting daily deviations from some safety specifications (see the concept of "normalization of deviance"). Her analysis also allowed her to draw attention to the "weak signal" concept, which is defined as information whose significance for the threat to a system's safety is not obvious. The notion of "weak signal" is similar to Turner's model of events not perceived by the organization during the incubation period of an accident. The accident was "being prepared" [GUI 11]. Fragmentary indices should be linked, like the pieces of a puzzle to make it possible to recognize a threat that can be reduced [LES 01, JOU 14]. Vaughan [VAU 97, VAU 99] was especially interested in questions on the side-effects of decision-making policies within organizations by a "trickle-down effect" and in questions on the "dark face of organizations" and "organizational misconduct", that is, in unfounded beliefs, in censoring and self-censoring, in bias in the decision-making process, in shortcuts and distortions in communications.

More recently, research on the notion of resilience (see section 6.6.6) [HOL 06] have extended some directions on the study of everyday reliability, but sometimes in more degraded conditions. Hollnagel [HOL 14] notes that when we think of industrial safety, it was generally with reference to its opposite, that is, the study of failure, that is, the absence of safety. The traditional vision of safety, which it calls Safety-I, has therefore been defined by the absence of accidents and incidents, or as "the absence of unacceptable risks". Consequently, research on safety and safety management would focus excessively on the risks of dysfunction of the system rather than on the conditions allowing safe operation (called Safety-II). Unlike the traditional vision, resilience engineering argues that "things will go badly" and "things will go well" for the same fundamental reasons. This corresponds to a vision of safety (Safety-II) that defines it as the ability to succeed in variable conditions. Understanding of day-to-day operation is therefore a condition necessary at the outset to understand an organization's safety performance [HOL 14].

Are French researchers still absent from reflections on industrial safety? In the domain of the nuclear safety of PWRs, we can note the contributions of Mathilde Bourrier on the study of maintenance who, following the HRO line of thinking, explains, among others that:

> organizational failures will only be understood if a better knowledge of the modes of collective action is developed within these high-risk systems, i.e. by focusing on the study of normal operation ([BOU 99], p. 3).

Similarly, Journé [JOU 99] studied daily operations by nuclear reactor operation teams, especially unexpected events resolution practices that rely on enquiries carried out by extended work groups and by articulation of different resources. Michel Llory extended the approach of regular analysis of series of accidents from different sectors [LLO 96, LLO 99]. He also argued that understanding of the accident occurs through analysis that goes beyond "operator error" and showed the importance of organizational phenomena in the occurrence of accidents, the relative lack of analysis of the work of operational staff and also of the role of engineers as well as of decision-makers, that is, of the whole organizational network of actors "involved" in risk management. Finally, he demonstrated that decision processes should form part of analysis as a potential cause for the occurrence of an event (see, for example, analysis of the Columbia Accident Investigation Board [CAI 03, STA 05]). These problems have also been developed in comparison with the near-miss accident at the American nuclear power plant at Davis-Besse [DOE 05].

Capitalization work undertaken by EDF R&D on the methodological approaches used in investigations after industrial accidents and in expertise at the IRSN gave rise to the development of formalizations of "organizational analysis" methods [LLO 10a] or "organizational diagnosis" [ROU 08b]. In addition, capitalization work on the lessons of accidents made it possible to develop knowledge on accidents [DEC 10]. Points of investigation are retained in the light of the recurrent, in-depth or root causes of industrial accidents, called "pathogenic organizational factors" [DIE 06] such as production pressures, organizational complexity, risk analysis and operation feedback failures, lack of staffing and competencies, a lack of demands from monitoring authorities, and so on.

To conclude this examination of research in the domain of industrial safety and nuclear safety, the following ideas can be retained:

– the concept of human error alone does not work;

– the occurrence (and recurrence) of event(s) is promoted by in-depth/root causes of organizational nature;

– an event does not happen at random; there is generally an incubation period during which the safety level degrades; this degradation, which is often detectable and detected, is an opportunity for prevention;

– decision and engineering processes are as important as technical failures and human errors in understanding the event;

– sustained attention to the daily work of each actor is a necessary condition for maintaining safety.

6.4. Human and social sciences methods for collecting and analyzing data

Human and organizational dimensions of risk management are thus a form of applied discipline that remains relatively new for high-risk industries. It would thus be more appropriate to stick to a set of practices, formed by HOF specialists contributing to risk prevention, which has been structured gradually since the 1980s, in nuclear and other industries. It makes sense to recall here that this profession is exercised by individuals with various profiles (experts, HOF relays in particular) with training in various human and social sciences and more or less in-depth actions in the matter of HOF. HOF relays are mainly engineers and technicians who are trained in some practices and in the use of frameworks. However, each manager, engineer or technician in a high-risk industry reflects and also acts on a daily basis on the human and organizational dimensions that involve the organization of work, engineering and risk management.

It makes sense to distinguish the roles and missions of HOF experts in the prevention of risks: some help engineers and managers to refine more effective, reliable and safe provisions and practices; others are responsible for evaluating and inspecting, or defining regulatory requirements. However, the approaches mobilized and the methodological benchmarks are shared.

HOF specialists are interested as a priority in the performance of the sociotechnical system and its real operation. When they participate in the design of a technical device or an organizational provision, their approach seeks to understand and assess, particularly through analysis of existing situations or simulations, concrete usages, direct and indirect effects, and potentially deviations from correct usage that could compromise risk management.

Subjects for analysis commonly entrusted to HOF specialists involve real practices implemented in particular work activities or situations that include nuclear safety challenges such as control room monitoring, a maintenance intervention, calculations by an engineer or a manager's decision. They focus their attention on collective work processes. They seek to characterize the effects of provisions and integrated management systems (production, safety, quality, environment, etc.) on practices. They consider managerial and cultural dimensions. Thus, levels of analysis combine microscopic levels at individual level, with collective, inter-organizational, and then macroscopic levels.

To carry out these types of analysis, HOF specialists collect various types of data, including formal and informal data, from several sources such as documents, observations and interviews. Analyses of documents aim to understand the rules and procedures in place, as well as information made available to those involved. They make it possible in particular to describe and analyze the formal system put in place to master risks. Observations of individual and collective work aim to collect traces of actors' real activity, their representations and the meaning they give to their actions. The HOF specialist will also spend time on interactions between actors, especially on informal exchanges [MIN 89] where points of view are discussed more openly. Indeed, we classically distinguish what is officially known, said "frontstage", [GOF 87] from what is said "backstage". Interviewers make it possible to collect explanations and rationalizations but also discourses and opinions. Informal and subjective data are often considered not to be very reliable by engineers. Nevertheless, to understand reality, to analyze situations with which actors have been confronted, it makes sense to collect the rationales behind actors' actions. Many methodological guides exist in the human and social sciences to conduct interviews and carry out observations. The network of actors to question changes scale when we move from an activity that is sensitive for safety challenges to a management process or the organization of a site and in this case, from a few interviews and observation days, analysis then requires several days of observation and dozens of interviews. Organizational diagnostics affecting safety management on several nuclear power plants have been the subject of dozens of observation days and more than a hundred interviews [ROU 08a, ROL 09, DEC 14, GAR 16].

Analysis then aims to compare these diverse sources of data and types of data (objective and subjective). Analysis is an objectivation exercise carried out by a HOF expert or more often by a collective enquiry [ROU 11]. The data are compared to standards, criteria, frameworks, models of the organization's human functioning, good practices, cases of accident and reliable organization. These data and their interpretation should be debated with the actors involved to evaluate it. More generally, analysis is carried out continuously and is an exercise of investigation [JOU 05, JOU 06], seeking evidence, indices which are similar to clinical approaches in medicine and psychology [LLO 18]. This multidimensional analysis requires varied skills [RAS 97]. Multidisciplinarity remains a challenge when addressing large organizations where the division of work separates professionals and disciplines. Practices of coordination and cooperation during these analyses are essential for achieving an integration of viewpoints and demands.

6.5. Making human activities reliable

One of the main missions entrusted to HOF specialists has been to explain and understand sources of "human error" and to find methods and devices to prevent them. Efforts undertaken sought to make human activities reliable by going beyond behavioral issues especially by focusing more broadly on the underlying conditions that influence the quality of the work in general and the management of risks in particular. From the "human factor" [DEJ 95], we have moved to "human factors" showing the multiplicity of dimensions to consider. In addition, it makes sense to consider the work of all the actors who contribute to managing risks and not to limit ourselves to actors at the bottom or end of the organizational chain.

6.5.1. *"Human error": man is a fallible reliability agent*

The TMI accident drew the attention of the nuclear industry to human errors in particular. One action by the operators transformed "a trivial operating incident into a serious accident" [KEM 99]. Humankind was thus perceived as a or the weakest point. A question was asked by some: can mankind be replaced? Automation and robotization, do not remove all errors, they shift them to the design, maintenance and monitoring stages, and today create new errors in the core and in opacity of algorithms.

On a daily basis, each person makes an effort to prevent errors. However, it happens that some are committed and corrected automatically [AMA 13]. Work has been carried out to highlight human error self-correction and recovery practices (e.g. the event database, *Recuperare* at the IRSN [MAT 02]). In short, "man is a fallible reliability agent" [FAV 70, MAZ 93]. In fact, according to the adage, *Errare humanum est*[9]! So what can we do? From the perspective of prevention, it is not reasonable to be satisfied with this fatalistic explanation in at-risk industries. We note that we often forget the second part of the adage *(sed) perseverare diabolicum*[10]. The questions that operational, prevention and reliability experts should ask themselves are as follows: What was the chain of activities that caused this error? How does the person involved explain it? If another person was placed in a comparable situation, could they be induced to repeat the same error? What lessons can be drawn from this?

This notion of "human error" when used alone, is in reality an intellectual trap that is dangerous for risk management. It may seem to provide a convenient explanation but it does not in reality provide the insights for understanding that are nevertheless indispensable for drawing lessons and identifying actions for progress.

9 To err is human.

10 (but) to perservere is diabolical.

It becomes reductive if one only attributes the result of a human performance to an intrinsic quality in the man that carried it out. By doing this, it then hems in, or indeed stops analysis at an explanation that appears sufficient (the "stop rule" [HOP 03, VAU 21]). Used as an explanatory factor, it thus prevents the search for an understanding that requires questioning about the conditions of the real activity, the individual's train of thought, the consideration of the effects of the working situation and the organizational context. In addition, it provides a judgment that benefits from wisdom obtained after the event [REA 90] and so introduces the risk of retrospective bias. Finally, we point out a classic attribution bias that consists of attributing an error to an individual's behavior or characteristics Paradoxically, when this affects us directly, we look for elements in the situation or context that could have led us into error [KOU 00, MBA 09].

Most fundamental from the perspective of preventing errors, one should not satisfy oneself with simplistic, ready-made explanations:

> Formal investigations into accidents generally start from the hypothesis that the operator must have made an error, and if an error can be attributed to him or her, this is the end of any serious investigation. The discovery that design faults were responsible would lead to the enormous costs of shutting down and starting up again; the discovery that the leadership were responsible would be a threat to those in charge, but the discovery that the operators were responsible preserves the system, with some injunctions for better training [PER 84].

In this sense, Dekker invites us to move to a new approach to human errors [DEK 06]. Efforts to establish a "work climate" or a "just culture" allowing the communication and discussion of the error are essential before any evaluation [REA 97, EDM 99, DEK 07]. In the opposite case, these errors are not indicated and may remain hidden, which sustains "organizational silence" [ROC 16]. From this point of view, companies have progressively made efforts in this direction on operation feedback (see section 6.6.7).

We return to the TMI accident. The human operators were accused of having stopped a nuclear safety automated system, that is, the injection of water into the primary circuit and their decision was based on an erroneous diagnostic of the situation. In reality, they faced multiple alarm signals in the control room ("the Christmas tree effect") and with ambiguous information. They developed an erroneous representation of the state of the reactor that led to them to generate a series of logical but inappropriate actions. The main reason was that the pressurizer relief valve that had received the order to close remained stuck in an open position, leaking, without the operators knowing. The synoptic in the control room indicated the order given but did not indicate the real position of the valve, that is, the operators believed it was closed. One of the fundamental lessons of the accident is that many design faults on man–machine interfaces lead the operators into error.

For the work of making human actions reliable, it is useful to distinguish different types of error to prevent [RAS 83, REA 90]: mistakes that appear at the moment of carrying out actions; lapses at the moment of executing action plans; these errors are associated with automation; there are other types of errors among which we find those linked to poor application of rules but also those linked to gaps in knowledge. There are communication and decision biases, which we will discussed in the following.

In the middle of the 2000s, six reliability practices were introduced in the nuclear sector in France. They are part of "human performance" approaches as those practiced in North America: the "pre-job briefing" aims to prepare the working team on the key stages and risks of intervention with the associated parades; the "one-minute's stop" invites the human operator not to hurry into action but to ask questions globally on the dynamic of the working situation and safety, especially in case of hazards; "secure communication" aims to check that the receiver has really understood the message; "cross-checking" allows a prior and independent control of actions undertaken by a human operator; "self-monitoring", is an individual act that can require following an instruction carefully and verbalizing their action(s); finally, "debriefing" aims to capitalize on the experience feedback of the operation. We note that these reliability practices enlarge the error factors, which are considered, since they deal with conditions linked to work situations for which these practices are even more justified: time pressure, urgency; too great (or too little) self-confidence; a habitual or routine activity or a lack of experience in the activity (novice worker); unclear roles and functions of those involved; a change in planning or the conditions of the intervention; a recent change in the procedure; a lengthy, laborious activity causing significant fatigue and high levels of stress; a complex activity involving multiple actors; frequent interruptions and a loss of vigilance [EDF 06].

This broadening of analytical focus is essential, as it means considering human actions in their changing contexts, activities in work situations where various variables or diverse resources (procedures, knowledge, expertise, etc.) must be used widely. Although reliability practices make it possible to reduce some risks of error, other factors that promote errors or their prevention should be covered.

6.5.2. *Training*

In 2018, EDF's French nuclear power plants were operated by nearly 23,000 employees, of which around 7,000 were dedicated to maintenance and around 25,000 were employees in 400 sub-contracted businesses. It is necessary to add

20,000 employees from engineering divisions of EDF and Framatome [COU 20]. Training needs were therefore immense and have been the subject of particular attention since the construction of nuclear power plants with American technology. Basic knowledge was formalized by EDF and made available to employees [EDF 16].

Requirements are very high in some professions such as for operating reactors, where the complete training of human operators, safety engineers and operating managers lasts around two years, by alternating theoretical expertise and practical know-how, especially with training on a full-scale control room simulator, as well as the realization of actions supervised in the power plant, such as lineage of circuit. Simulator training includes day-to-day actions, but also the hazard management and the implementation of incident and accident management operating procedures [COU 20]. Some elements of Crew Resource Management [FOR 16] training tested by the aviation sector could enrich training on the means and biases of communication within work teams.

It should be noted that training only provides a promise of skills [ROU 11]. After training has been delivered to be able to carry out important activities for safety, clearance is granted. This is a managerial and administrative act of recognizing the qualification required. The training course and clearance are recorded in an individual training or professional training record that is subject to checks by the Safety Authority. However, it is important always to ensure a sufficient level of experience for these operators. Thus, since 2004, progressive, more targeted, clearances were put in place to better meet the needs and constraints of professional training courses [COU 20].

More generally, EDF [COU 20] implemented generic training for all new recruits with a dedicated course of around three months. Many professions have implemented specific training. For example, operator training simulators, the maintenance professions (electricity, automated machinery, mechanics, boiler-making, etc.), have put in place training sites to manipulate equipment. They are partly open to contractors and allow training before an unusual intervention or that is exposed to high radiation. Complementing training, there are work situations where practice can be supervised by a peer.

In addition to the initial acquisition of skills, skills' maintenance requires regular practice, and indeed repetition, or the person runs the risk of losing their abilities, especially for control operators. Finally, new skills should be acquired following technical modifications, and technological and organizational changes.

Similarly, the Fukushima accident led nuclear operators of nuclear power plants to develop teaching and practical training to develop abilities in adaptation, resilience and stress management in extreme situations, especially for members of the Nuclear Rapid Action Force, and the crisis and operations teams [ALE 18]. Some training methods used in civil protection for crisis management or by the army were transposed. Others rely on training to debriefing methods and on the use of operation feedback from crises such as the 1999 storm that impacted the nuclear power plant at Blayais, or natural disasters impacting the EDF electricity grid.

Beyond a detailed analysis of specific training needs, ergonomic analysis of work situations and adult learning invite us to enter into discussion with peers and ergonomics analysts and trainers about "tricks of the trade", about the artifacts used and operational and pragmatic concepts [VID 07, FLA 19]. This is the price at which training will bring us closer to the skills required in work situations (e.g. expertise, know-how) or prudence knowhow [CRU 83].

6.5.3. *Applying the procedure or demonstrating skills?*

The utility and use of operation procedures whether in a normal situation or in an incident situation are not in doubt for anyone. They are designed to help operators respond to the situations they face [DIE 98]. Efforts to put these in documentary form have been made so that they can be read effectively and clearly in emergencies.

A fundamental question concerns the difference between the vision that the designers have of the role of the operators and the vision that the operators have of their own role. For the designers, the operators should be guided and helped and this means there is a need for strict compliance of the procedure. For the operators, their role is both to follow the procedure but also to respond to situations that do not correspond "exactly" to the procedure (failing sensor, missing or erroneous information, etc.).

We must be aware that no procedure can be wholly exhaustive [DIE 98, BOU 99] – the exhaustiveness of procedures is therefore a myth – unless they lead to a "combinatorial boom" and become inapplicable as a result due to the multiplicity of possible paths. The application of procedures is not therefore a guarantee of absolute safety. We thus expect an "intelligent application" of procedures in a given situation, which therefore depends on skills, experience and professionalism. It therefore seems that the questions to solve are as follows: What is the level of skill required for users of this procedure? How do we define and attain this level (which depends very much on the situation to be managed)? What margins of maneuver are left to the operator (how much right do they have to take the initiative)? Who takes responsibility in "case of failure"? If clear answers to these questions are no guarantee of the absence of dysfunction, they at least ensure a clarity of responsibilities, thus enabling a path in the direction of better risk management.

Many activities that contribute to reliability and nuclear safety are not proceduralized and could not be: a great deal of daily work, movements between two work stations, professionalism rely on the need for procedures but generally exceed them. The total knowledge, expertise and know-how is not codified and a significant part will remain impossible to codify, difficult to make explicit or consciously known, and therefore tacit [ECH 18]. In addition to approaches documenting the tasks to be carried out, it is necessary to rely on the consolidation, perenniality and density of skills. The problem of skills has involved several HSS disciplines for decades. Rules should be considered as one resource among others to be mobilized to act.

Moreover, a study [BOU 99] showed that conformity to rules depended less on national culture (in this case French, expected to be more flexible than American culture expected to be more compliant) than on the power delegated to modify rules given to users as well as the organization in support of their adaptation.

In France, in high-risk industries, exchanges in the last decade re-address the contribution and limits of rules, especially through the balance between an approach relying on rules and one relying on professional management delegated to the actors in charge [BIE 13, FUC 13, BES 17, COF 19]. Operations activities in nuclear power plants are characterized especially by the occurrence of hazards and unforeseen events that will mobilize adaptation, problem resolution and innovation skills. Consequently, it is also a matter of reinforcing attention to the constitution of these individual and collective skills [LAR 08].

6.5.4. *Analyzing real activity and work situations*

One of the contributions of ergonomics has been to demonstrate the difference between the notions of task and of activity. The task corresponds to prescribed work, that is, the definition of objectives to be attained in conditions set and decided in advance by individuals outside the activity. The activity corresponds to real work, that is, to the implementation by the operator of their abilities to carry out the task. The activity depends as much on the individual as on the conditions encountered in real time. This fundamental difference means that the design of operators' tasks should integrate their actual activity. In other words, knowledge of their activity (their actual work) depending on work situations is the main condition for designing tasks.

It is therefore recommended to analyze the actual work of operators to adapt tasks to the conditions that the latter may encounter. In the context of "totally" new tasks, it is always possible to consider similar or related tasks, that is, to analyze the work of individuals carrying out this work and to reflect to transfer the results shown to the future work situation. These recommendations mean that the tasks are not only designed in a design office but are also sustained by analysis of the actual activity of operators on the ground.

Methodologies resulting from ergonomics make it possible to carry out this type of analysis and have also been used regularly in the nuclear industry since the 1980s, especially on critical activities such as control (e.g. lineages), or on lines of defense (e.g. control of activities) [COU 20]. But other disciplines in the human and social sciences, psychology, sociology (of activity, of work, of sciences and technology), anthropology or management sciences make it possible to enrich perspectives brought to the activities, practices and work of operators, as well as engineers or scientists. With the sociology of work as a starting point, the subjective and physical dimension of activity and work on machines is underlined [BÖH 98]. In the framework of theories on organization, attention to activity and to practices should be strengthened by broadening the focus of observations, to focus on the different dimensions of a work situation, considering the organization of the work, as well as history, habits, routines, individual and collective regulations, material aspects (e.g. artifacts, tools), and professional identity and communities of practice [GHE 19].

6.5.5. *Man–machine interfaces: the case of control rooms*

The TMI accident triggered numerous studies on the usability of man–machine interfaces. The control room is the place where operators and their supervisors have available all the information and control devices allowing them to pilot the nuclear reaction process. The design hypothesis is therefore to provide operators with all the information on the state of the system, the equipment needed for them to operate it and giving them access to all the required commands of devices (e.g. valves, pumps). The TMI accident showed that this objective is necessary but demanding for designers.

First of all, information should be presented so as to correspond with the working image the operators have [OCH 78] (e.g. synoptic information should correspond to the position of the devices and not to the order a device might have received). Then, the information provided should correspond to the reactor's situation (e.g. ordering alarms to avoid the "Christmas tree" effect). Information means should be paired visually with action means so that operators immediately see the result of their actions.

With these lessons and to benefit from new possibilities in information technology (to create a hierarchy of the information and alarms, to display different visual representations) [COU 17], EDF decided, at the start of the 1980s, to design a totally computerized control room for reactors N4[11] (1450 MWe), without changing the level of automation and without changing the general organization of the operation team. To face the challenges that this new control room represented, EDF took a number of decisions, among them: integration of HF specialists into the design team and building a full-scale simulator to test and evaluate design choices. It must be noted that the design team, like the evaluation team was multidisciplinary (HF specialists, operation specialists, designers) [LEG 98] and mobilized safety authorities [TAS 00]. This approach (integration of HF specialists and use of a simulator) was used again for designing EPR reactors [COU 17].

It can be noted that although the design of the control room has been validated by stakeholders, studies carried out have also shown that the level of computerization cannot be disconnected from the level of automation; the design of control room images should consider the mnemonic load; and operators' needs for coordination should be considered in the overall work station architecture.

6.5.6. *Consideration of HOFs during design and modifications*

The design or modification of technical and material provisions (e.g. documents, tools, equipment), by implementing a user-centered HOF approach, aims to better adapt these provisions to the constraints and needs of the "human at work". They also have effects on work situations which it makes sense to anticipate (e.g. the effects of computerization). These approaches seek to consider the existing situation to specify the design objectives with regard to operational needs (e.g. to create a hierarchy of information) [COU 20]. Nevertheless, the existing situation can sometimes only be approximated; in this case, "reference situations" are identified and serve as a basis for reflection. Points of reference are thus drawn in operation feedback from the design of previous devices, events or other elements of knowledge. Within the design process, the development of devices can refer in particular to many international standards (ISO or IEC) or nuclear bodies, especially American ones for the consideration of HF in engineering processes. We note that there is a control authority guide [ASN 17] on the design of pressurized water reactors (PWRs) that integrates some of these principles. Finally, it is essential to check that the implementation of devices makes it possible to meet needs effectively. Along design process, recourse to models or prototypes is thus essential. In this context, test campaigns have regularly been carried out on full-scale control room simulators: for example, in the years 1994–1995 for the use of computerized instructions at the plant and to understand their impact in particular, such as the effects of reading a step-by-step guide, framing effects, or indeed a passive follow-up of instructions [COU 17]. This design process should be considered as

11 There are two pairs of such N4 reactors in France (Chooz, Civaux).

iterative, especially due to evolutions in design choices over the course of these devices' development. An HOF engineering program should be integrated into the structure of the organization in charge and from the start of a system's design process.

These design situations remain relatively limited in number and have historically been created for control rooms. In contrast, nuclear power plants implement modifications each year (a hundred or so, even, during 10-year inspections) especially to integrate generic operation feedback drawn from lessons from events that have happened on the nuclear reactor fleet. High-risk industries have had their share of accidents of which one of the causes was linked to the effects of a modification (e.g. the accident at Flixborough in 1974 in the United Kingdom [MON 13]). High-risk industries have for some decades regulated the installation of modifications, especially those considered to be notable, in view of the risks that they can introduce [OCD 09]. Since 2007, EDF has put in place an SOH approach to considering social, organizational and human aspects for designing modifications [LEG 12]. The objective is to anticipate the effects caused by modifications on work situations and practices and to characterize risks. In the "strategic phase" of the method, this means marking, during a "sensitivity analysis", the modifications most likely to have effects and to entail risks. This phase serves as a sorting process and hierarchization, and aims to proportion the implementation of the SOH approach to the challenges presented. Engineers are regularly trained and benefit from guides who provide aids to questioning and methodological references, but fragilities in the exhaustivity and effectiveness of the process are sometimes observed by safety authorities [COU 20]. This type of approach remains all the more necessary in view of the design of new nuclear reactors technologies and for the implementation of new technologies.

6.5.7. *Operation actions and their feasibility*

Since operation of energy produced by nuclear reactors began, the organization of the operation team has evolved more than once. The TMI accident [KEM 79, ROG 80, LLO 99, COU 20] showed all the benefit in having a safety engineer to evaluate independently the conformity to safety rules of operating activities. This engineer was first added to operation teams, then left the shift operation teams during reform in the 1990s so that they would always have "fresh eyes" with regard to the situation.

In a control room, two operators and possibly their supervisor ensure the coordination of operating activities (fluid transfers, stopping and starting of pumps, etc.) from the control room in interaction with consignment officers and agents who carry out local actions (opening/closing of valves, rounds to detect leaks). They also ensure monitoring of the state of the nuclear reaction and of the system. Far from being a passive activity, many studies on the operation and monitoring of continuous processes in the 1980s showed the importance of anticipation, training, experience and the focused and global character of the monitoring, but also of sensitivity to serenity in the control room, especially for the coordination of monitoring [COU 17]. After reflections in 2006 for reinforcing monitoring in the control room in normal operation and after the Fukushima accident for the reinforcement of supervision in an extreme situation, EDF decided in the mid 2010s to add a "head of reactor" operator for each reactor, so as to have, from 2020 (the time to recruit operators, to train them and allow them to acquire experience) a supervisor present permanently in the control room beside the two operators. The goal is to have one operator keeping their head above water [COU 20].

At the Fukushima Daiichi nuclear power plant, the operation and crisis teams were confronted with unforeseen situations with much degraded intervention conditions because of the significant impacts of the tsunami on the site [GIS 15]. In France, on the basis of 2011 complementary safety assessment (EU "stress tests"), the Safety Authority asked EDF to identify the human actions required the management of extreme situations following out-of-the-ordinary natural aggressions, which could lead to the loss of electrical sources or sources of cooling. In addition, personnel should be qualified to intervene in these situations and in sufficient number [ASN 12]. To reinforce the robustness and resilience of its control and crisis teams, EDF led tests on operation control room simulators from 2014 to 2016. In addition, a review of operating procedures for situations where there is a loss of electrical and cooling sources made it possible to identify the importance of prioritizing essential actions that it makes sense to carry out to avoid core meltdown. To consider extreme conditions for intervention and the risks to which those intervening can be confronted, an approach has been developed by the operator to check feasibility. Making these actions feasible in an extreme situation may in some cases necessitate some modifications of the installation (to automate and control it remotely) or to install protection inside or outside its buildings [COU 20].

6.5.8. *Quantitative approach to human reliability*

For the needs of probabilistic safety studies (see Chapter 3), probabilistic studies on human reliability (PSHR) have been developed by EDF and by the IRSN independently and their updates are regularly compared. PSHR modeling is focused on the probabilities of failure of actions and operation missions, especially diagnostic and execution errors. Today, there are tens of models and quantitative modeling remains a delicate art [COU 20] without forgetting that qualitative approaches sometimes remain heterogeneous and dependent on experts and their methods, although operation feedback data make representative values available. We can mention Swain's THERP model included in the NUREG 1278 report from

1983. EDF and the IRSN have developed their own models (respectively MERMOS and PANAME). Data collected in France and internationally during test campaigns on simulators are used to provide probabilities that are the subject of debate during scientific conferences. For reliability rate calculations, it should be noted that data listing failures is more reliable than data listing situations where an error occurred in daily activities, which are more difficult to trace. For the modeling, some of this data are combined with expert judgment (especially the testimony of agents in the operations team, but also HF experts). For the MERMOS method, the scale of estimations lies between "very improbable" for which the probability is 0.01 and "very probable" for which the probability is 0.9 [COU 20].

After the TMI accident, a four-part EDF-Framatome-Westinghouse-IPSN study program had the aim of studying operators' response time in accident situations. The IRSN (IPSN at that time) carried out research from 1996 to 1998 at HAMMLAB on the Halden reactor in Norway on the impact of automation on operators' performances [COU 17]. Independent variables were assessed by HF experts such as the team's and operators' overall performances over windows of time; measurements of awareness of the situation and confidence in automated machines were collected from answers to questionnaires during or at the end of the test.

6.5.9. *HF in maintenance interventions*

Concerns were also raised about maintenance interventions likely to generate "latent faults" that "reside" in the system like "pathogens", before they are involved in an accidental sequence (see the "Swiss cheese" accident model [REA 90] described above). In France, three incidents in 1989 during maintenance interventions triggered in-depth investigations by EDF (see the Noc report [COU 20]), leading to evolutions in organization and to monitoring by supervisory authorities.

Ergonomists from IPSN led work observations of about 50 workers during reactor' outages in 1992 and 1993. The study revealed insufficient consideration of the conditions in which interventions were carried out [COU 17] during preparation and operation feedback where difficulties in intervention were not sufficiently relayed to preparers and organizers. Indeed, workers encountered multiple hazards and had to resolve them by mobilizing technical support resources or by adapting to the situation professionally. To continue these reflections and exchanges between EDF, the IPSN and the CNRS, a research by Bourrier [BOU 99] would be produced on four power stations (two French, two American). They would illustrate four different organizational reliability regimes in relation especially to responsibilities and resources for modifying rules (respectively for American nuclear power plants in 1992-1993, delegation to responders or technical support from an engineer; and for the French ones, decoupling between practices and rules internally in 1991 or stricter respect [for rules] by contractors in 1994).

These difficulties gathering intervention conditions as planned would be observed again in 2012–2014 during IRSN evaluations [COU 20] despite the implementation of meetings dedicated to check that the prerequisites conditions had been brought together to allow interventions. Provisions would be reinforced by EDF, such as logistical support with the implementation of a “a one-stop shop”, a contracted supervision ratio of intervenors, and the association of contractors at the preparation stage. Equally, the articulation between the regulated and planned, on the one hand, and the managed and unexpected, on the other hand, is a significant work axis to make the organization more robust with a more formalized capacity and also with more explicit procedures to manage waiving if necessary.

Monitoring of contractors’ interventions has been imposed since 1984 (Decree on Quality). It should enable to check globally that the safety policy is respected but above all that the compliance to the demands defined for some activities and equipment. Monitoring actions should be proportionate to risks. For example from 2013, EDF raised the requirement on the technical skills of controllers, but also on the size of teams and strengthened some of these provisions [GAR 16]. The operator also ensured a “re-internalization” of some maintenance interventions to maintain the skills of controllers.

6.6. Making the organization of work and risk management reliable

The shock of the accident at Chernobyl in 1986 aroused new worries about other key elements of risk management, such as the role of hierarchies in decision-making and the priority given to safety, the safety culture of personnel and directors, but also the controls ensured by safety authorities. The Fukushima accident in 2011 confirmed the need for these fields of analysis. However, the extension of analyses from HF to organizational and social factors was gradual and has still not been achieved. The 2000s were marked by the gradual appropriation of the organizational paradigm for questioning organizational reliability and safety management. Analysis of provision for control and governance of risks was strengthened after the Fukushima accident.

The fields open to analyses are multiplying at different levels of the sociotechnical system (sometimes called micro-meso- and macro-scopic) to better consider the collective dimensions of the organization of work but also of the determining role of management and control authorities. It is also a question of considering: informal dimensions, the circulation of information, conditions and provisions for coordination especially when preparing activities, cooperation practices, the management of human resources, systems and processes for risk management (especially risk analysis, operating experience feedback, change management, crisis management, etc.), engineering and regulation processes, relations with sub-contractors and stakeholders, organizational and groups cultures, power games, conflicts and controversies from a socio-historical perspective, etc.

6.6.1. *Quality approach and safety management systems*

Although the beginnings of the nuclear industry rested essentially on multiple technical and professional excellence, quality assurance and then quality approaches have been gradually imposed in many industries since the 1970s and 1980s, with some methods from Japanese industry as a model. Quality, safety and environment management systems were deployed from the 1990s and integrated management systems since the 2010s. In view of the importance attributed to these management devices, they have been enshrined in regulations.

In 1984, the "Quality Decree" of August 10th formalized in France a number of requirements on the quality of design, construction and the operation of basic nuclear installations (with notably, for equipment important for safety, the notions of defined requirements (DR) for equipment important for safety and activities affected by quality (AAQ)). These requirements would involve, among other subjects interesting to HOFs such as organization, human and technical means, skills (training, qualification, authorization of personnel), checks on AAQ, verification, the monitoring of contractors, document management and the handling of anomalies and incidents.

Until 2012, the text of this Decree was the regulatory basis that permitted inspectors from the Safety Authority, with the support of HOF specialists from the IRSN, to carry out inspections on these different subjects. These regulatory provisions have been reviewed and broadened by the "INB Decree" of February 7, 2012; the text imposes the implementation of an "integrated management system", it set the requirements applicable to activities important for the "protection of protected interests" (as defined in the environment law[12]) and requirements on the monitoring of external responders, as well as requirements on the consideration of HOFs on analysis of significant events. Additions were made to requirements on sub-contraction by decree no. 2016-846 from June 28, 2016.

Important subjects are management system processes (e.g. risk analysis, operating experience feedback, purchases, and so on). In their entirety, these processes require the implementation of provisions for organizing work,

12 In France, "code de l'environnement".

communication and decision-making on different subjects. Nevertheless, although they guarantee a minimum of coordination, it remains the responsibility of actors to commit to safety and especially of managers to make good use of these provisions for open discussions and in-depth debates on technical, human and organizational aspects. In the contrary instance, these may become an "empty shell". The integrated management system has the aim of coordinating, making coherent, or indeed "decompartmentalizing" these different processes. Indeed, it makes sense to recall that the effects of these processes are materializing in each work situation. In these situations, the approach is in essence multidimensional and stands out from administrative management, which relies excessively on "siloed" approaches. Consequently, the difficulty for actors is to articulate the different requirements in activities and if necessary make arbitrations between priorities. This challenge is a daily one and an immense one for operators; it depends on the individual and collective resources allocated to them and the upstream processes mentioned above.

The nuclear industry is also characterized by several independent internal and external control loops. Workers (sometimes a contractor) are monitored technically by a peer (from their business, then by a controller from the operating company). The operation performance is monitored every day by the independent safety department engineers. Quality auditors from a power plant audit management processes and interventions. External auditors and inspectors from the EDF nuclear fleet central division, especially peers from other sites, compare practices and conformity to rules. Auditors from other nuclear operators carry out "peer reviews" within the WANO association or international inspections of the IAEA (e.g. OSARTs and SCARTs for safety culture).

6.6.2. *Safety culture*

The disaster at Chernobyl in April 1986 led the IAEA to choose, as a lever for the prevention of nuclear accidents, the notion of safety culture, of which it gives the following definition in a report currently called INSAG-4:

> Safety Culture is that assembly of characteristics and attitudes in organizations and individuals which establishes that, as an overriding priority, nuclear plant safety issues receive the attention warranted by their significance ([IAE 91], p. 5).

This definition states the primacy of safety in the operation of nuclear installations. It advocates also that vigilance should be exercised at all levels: operators, management and control authorities. The INSAG-4 determines the characteristics that make it possible to obtain and maintain safety culture:

– For individuals, awareness of the importance of safety, competence and knowledge retained though external training and independent training.

– For management, commitment demonstrated by the priority accorded to safety. Management also ensures that everyone is motivated through its leadership, through defining objectives and through a system of rewards and sanctions. Management is also in charge of supervision with, in particular, a capacity to respond to questions from personnel. Finally, it ensures the definition of tasks to be carried out, and each person's roles.

The fundamental contribution of the INSAG-4 is to encourage a questioning attitude at all levels and all times, i.e. an attitude that is both rigorous and cautious to ensure safety management. Operators' responsibility for safety is considered paramount. Regulatory instruments and the resources attributed to regulatory bodies should be sufficient to make it possible to make safety a daily preoccupation, without unwanted interference from the state or operators ([IAE 91], p. 18).

The concept of "safety culture" is deployed in many industries [GUL 00] but has continued to pose operationalization issues despite efforts to develop methods using questionnaires. Although this concept has made it possible to place the question of HOFs on the agenda of engineers and directors [LAG 11], it has also had perverse effects such as halting questioning in event analyses for an explanation that appears satisfactory "a lack of safety culture" [ROU 11]. The excessively instrumental and behavioral character and the limits of measurement continue to pose problems [FAS 11]. Many researchers have been interested for the past two decades in some work in anthropology and sociology [FUC 12, SAN 17]. Bourrier [BOU 99] rejected the cultural hypothesis (in this case French or American) in the explanation of compliance to rules in favor of greater weight placed on the work organization and on the power to modify rules. Discussions during the IAEA conference in 2016 converged on the idea that it was preferable to evaluate the effects of organizational culture on risk management rather than the safety culture as such [IAE 16, BES 17, GIS 16].

6.6.3. *Forward planning of skills and workforce – human resources management*

High technology industries such as the nuclear industry mobilize a multitude of professions with advanced skills, whether internally or with designers and manufacturers of equipment or sub-contractors. The rise in skills has been occurring over many years and skills density is essential for it to promote "critical sizes", the transmission of skills, the sharing of experience, redundancy and cross-checks. From a rationale based around training, human resources departments have wished to direct themselves toward a rationale based around skills, especially since the 2000s with the implementation of forward planning of skills and workforce.

One of the contributions of this type of approach is its quantitative and predictive character of 2–5 years, and this was particularly necessary at a moment where the age pyramid of EDF agents indicated that, during about a decade from 2010 to 2020, nearly 50% of employees would retire in some areas [COU 20]. Indeed, this involved the generation of builders that had accumulated construction experience, and experience of power plant start-up tests and the earliest incidents. The nuclear operator implemented after a needs analysis, recruitment actions, training schemes, careers management and actions to acquire know-how. This approach relies on the identification of skills for different professions, with the help of observations in work situations and interviews, which are formalized in a "skills map". This tool makes it possible to facilitate the control by management of career paths, recruitment actions, covering between incoming and leaving workers during job changes, the establishment of succession tables, movement of personnel, the sizing of "skills incubators" [COU 20]. This process and this tool are subject to inspections by the safety authorities. Nevertheless, despite the implementation of such a process, particular attention should be extended to the duration (over several years) of the rise in skills; similarly, numerous recruitments can unbalance the practice of companionship [DEC 14] or overload training provision. In terms of perspectives, the challenges posed by the notions of individual and collective skills [LAR 08], communities of practice [LAV 91] and the maintenance of engineering skills [GAR 14] would merit new efforts in terms of study and operationalization.

With the extension of the operating life of reactors beyond the forty years envisaged when they were designed, substantial maintenance work should be carried out. At the start of the 2010s, EDF presented sub-contractors with multi-year forecasts to increase the volumes of work and extended the durations of sub-contractors' contracts to five or even seven years to allow sub-contractors to invest in recruitment and training. The Minister for Education was requested by the nuclear industry to provide professional forecasts of skilled roles in short supply in France (such as welders), and to put in place apprenticeship contracts [COU 20].

6.6.4. *Managing safety on a daily basis and decision-making*

The chief operating officer who supervises the shift operation team should include in decision-making all challenges (safety, radioprotection, performance, etc.) and associated criteria, as well as each operating and maintenance agent. After the accident at TMI, radioprotection safety engineers were recruited and trained on operating rules and technical operating specifications and normal incident and accidental operation. They carry out daily checks independently and have an autonomous vision focused on the safety parameters in incident or accident situations. This redundant line of defense is characteristic of the nuclear industry, and it has therefore been the subject of regular attention since the 1980s. This structural effort also makes it possible to provide technical support and advice on the interpretation of safety rules especially in the case of unforeseen issues. The outage project teams have also benefited from the appointment of an additional safety engineer within the team since the 2010s. High stakes decisions involve operation, maintenance, and safety departments as well as the power plant management. Arbitrations on critical questions that balance production and safety are drawn up and are the subject of checks by the Safety Authority. This sometimes calls these arbitrations into question, which causes changes of priority in the operation of reactors [COU 20].

In addition, safety is built up on a daily basis in a more distributed process, that is, it involves and is the responsibility of all actors. Day-to-day decisions are the subject of technical exchanges and points of view are regularly exchanged, especially for addressing hazards on a daily basis [JOU 99, DEC 14], with decisions that can move up the levels of hierarchy as far to national engineering and management headquarters [STO 10]. Decision criteria are multiple and evidence or elements of "proof" of safety or risk are mobilized to justify the decision and are traced. Doubts can also be exchanged [VAU 96, EYD 17] and should be subject to investigations. It is up to operators and HOF specialists to bring particular attention to: micro-decisions in actors' daily work; the quality of interactions; as well as individual and collective construction of the sense made of the situation ("sensemaking" [WEI 95]); the priority given to safety depending on the state of reactor process, the work situation and the organizational context [ROU 08a] and with regard to "production pressures" [MON 06]; decision-making rationales and heuristics, but also considering representation, perception, communication and decision biases [AMA 13, MER 16]. As an example, it is possible to cite phenomena linked to the collective permitting of daily deviations from safety standards [VAU 96], risks of confusion between reliability and safety [LLO 06], but also difficulties to represent the low probabilities of occurrence.

6.6.5. *Risk analysis, anticipation*

Anticipating risks is one of the main principles on which risk management relies. To this effect, risk analysis is thus one of the key provisions of risk management systems, whether in design or in operation. The construction of risk management is based on the hypotheses and conclusions chosen during risk analysis. The measures retained are proportionate to the risks that have been identified and characterized. Risk analyses should periodically be updated in view of evolutions in the system, context and knowledge, especially those drawn from operation feedback. Risk analysis was considered by the operator as a "lever" of safety management at the end of the 1990s, then as a lever for the overall performance of interventions and since 2014 has covered all protected interests of the kind in the environment law. If this evolution in doctrine is of a kind to promote an integrated approach to all risks, it could also lead to the dilution of the part relating to nuclear safety, which is subject to supervision by the operator [COU 20].

Risk analyses are carried out daily by nuclear operators. Their formalization is variable, as well as their exhaustivity. Many tools have been made available [DEL 13, ESR 20]; these are more or less well-used or transposed or depend on experts, their approaches and the means allocated to them [HOU 02, MER 10]. Methodological work has been carried out to better integrate human and organizational dimensions [ABR 99, LEG 09]. In the context of research in the domain of radiotherapy, FMECA (Failure Mode, Effects & Criticality Analysis) was revealed to be inappropriate for questioning the risks linked to activities and the organization of work, especially factors that weakened provision for risk management [THE 17].

In a risk identification process, composition of the work team is crucial and should aim to involve those preparing procedures, front line workers such as intervenors, engineers who have experience of the field and of procedures, "pillars of experience" [LLO 96] but also novices and external individuals who can spontaneously re-question collective standards, and the effects of precedence and jurisprudence [VAU 96], analysts and reliability specialists who mobilize operating experience feedback from other power plants or indeed other countries. A great deal of research in ergonomics and sociology since the 2010s has focused on the contribution of discussion spaces to debate work, practices and associated risks [ROC 16, CAS 16, THE 17]. Indeed, the representation of risks is socially constructed and is the subject of negotiations [CLA 99, MAG 13, POW 16]. Consequently, as risk analysis has a key role in framing and sizing risk management, risk analysis should therefore be the subject of an objectivation process [ROU 11], or subject to the organization of scientific dispute, that is, debates and contradictory dialogues. Indeed, although there are factual elements on proven risks that therefore draw consensus, several interpretations are possible for anticipated risks.

During operation, risk analysis is not conducted only in preparation for interventions. It is a dynamic process that it makes sense to update and which benefits from iteration. For example, gathering of prerequisites before a maintenance intervention should make it possible to check that the intervention conditions are met and where applicable, the risk analysis and countermeasures must be adjusted [COU 20]. Similarly, before an intervention is launched at the "pre-job briefing" and the "one minute-stop" when starting the intervention, checks should be made on the dynamic of the situation (e.g. change of plan, planning pressure, new coactivity, and so on). Finally, attention and vigilance should make it possible to detect events and deviations that may signal a dysfunction of risk management. Consequently, this calls into question the "situational awareness" that relies in particular on the appropriation of risk analysis by intervenors especially during the "pre-job briefing".

6.6.6. *Adaptation, resilience, emergency and crisis*

Management of emergencies was highlighted as one of the major axes for reinforcement after the TMI accident [LLO 99, COU 20]. Many provisions were designed and implemented after the accidents at Chernobyl and Fukushima (e.g. internal to the plant and managed by company and external emergency plan managed by public authorities, premises for a crisis management center). Crisis preparation [LAG 91, ROU 03] is the subject of provision for education and training [IRS 18], especially with crisis exercises based on particular scenarios [COU 20]. Those primarily involved are the EDF crisis teams (at sites, at national level, and at the FARN[13]) and the safety authorities (ASN, IRSN), as well as operations teams which include thousands of people in France. Some of these education and training provisions have been developed by relying on real crisis situations such as one that occurred at the Blayais power plant in 1999 (partial flooding of the site during a storm with high tide). The Fukushima accident highlighted the need to better prepare for unforeseen and extreme situations such as out of the ordinary natural disasters that tip organizations into breaking points and acute crises [COF 17b, IRS 21]. Although typical situations are envisaged in emergency plans, especially the national nuclear crisis management plan, the example of an unanticipated scenario is also identified [SGD 14].

To meet this type of need, the concept of resilience has been mobilized for a decade in the nuclear industry and the Covid-19 crisis underlined the benefit of this once again, especially in line with the concept of "business continuity". Resilience may be defined as the

> intrinsic ability of an organization to adjust its functioning prior to, during or following changes and disturbances, so that it can sustain required operations under both expected and unexpected conditions ([HOL 06], p. 16).

Work on operationalization of the concept of organizational resilience is still underway. This work aims especially at developing more adaptable organizational processes upstream, but also more generic individual and collective skills promoting communication and coordination or even meta-rules adaptable to the circumstances. The limit being that part of resilience or adaptation will depend on the circumstances and will be improvised in new situations [WIL 88]. Although variability and

13 Nuclear Rapid Action Force.

adaptation are constant realities of daily work as ergonomists have described them, resilience situations would correspond rather to break-down situations, considering the fact that perception, sensitivity and vulnerability to a degraded situation are dependent on individuals and organizations. Although reflections on resilience provide additional questions and perspectives on strengthening crisis management but also safety management, a critical view should be maintained on the possible transfers of responsibility to which it may lead, involving agents who will simply have to adapt, on a risk of renunciation to risk prevention and to investigate and address the root causes of failures.

6.6.7. *Event analysis and the operating experience feedback process*

In the French nuclear industry, attention to and regulations for operating experience feedback (OEF) from accidents, then incidents has been observed since the 1960s at the CEA [COU 20]. The first incident folders were formed in the 1970s. In the domain of producing electricity of nuclear origin, event analysis were undertaken following accidents at Saint-Laurent-des-Eaux in 1969 and 1980 on natural uranium graphite-gas technology reactors: they highlighted the benefit of foreign OEF with the occurrence of accident early warning signs (in the United Kingdom and in Spain) but also a form of forgetting, especially due to the deployment of new PWR technology [MAN 20]. In 1979, the TMI accident triggered the recognition of OEF as a central approach in nuclear safety. The accidents at Chernobyl and Fukushima reinforced the need for better use of OEF at international level despite the existence of provisions for exchanges that had been in place for decades.

OEF processes have evolved in their objectives, boundaries, organization, management, methods, and analysis tools (e.g. event analysis method, evolutions in databases). Since the 2010s, research and development projects have focused on the contributions and limits of big data and natural language processing (NLP) [IMD 13a, IMD 13b, DEC 18, ROU 18] and were therefore in the process of being operationalized in 2020 (see Chapter 7). From OEF-oriented analyses to the design of equipment, methods focus on OEF for using a system that is gradually being considered sociotechnical. The objective of OEF is to avoid the recurrence of similar events. This therefore leads to a search for generic lessons. This means analyzing what has happened and identifying the circumstances that promoted the event; then, the objective is to seek to characterize the underlying factors that explain the occurrence of the event; finally, it means understanding the reasoning and intentions of the actors involved in the events, but also prevention failures. It also means analyzing the potential consequences of the event ("what if", what would have happened if other barriers had failed). The lessons learned, from an event or series of events, should make it possible to develop knowledge, to share it to facilitate its use every day in decision-making or in defining corrective or progress actions that involve the reliability of equipment (see Chapter 2), organizations, or professional practices [ROU 14]. OEF is a relatively dynamic process, which is deployed on very variable time loops and should take better account of user needs [DEC 16].

Several studies of OEF practices in normal functioning and industrial accidents have shown that OEF processes showed limits and experienced failures with the recurrence of similar events [DEC 08]. In addition, there were barriers in learning [MAR 15] and margins of progress in better considering HOFs in OEF [MAR 14, BLA 16, BRI 16]. In 2009, EDF carried out an internal review of its OEF process, which showed the potential to learn from events was not sufficiently exploited [COU 20]. The operator then reinforced its human means by creating new roles for actors in charge of OEF in the context of a new process inspired by North American methods, the corrective action program (CAP). CAP relies on organization that prioritizes event analyses on a weekly basis that will be subject to monitoring through trend analysis of a more in-depth analysis. CAP makes it possible to process several hundred events per reactor per year, in addition to the tens of events that are the subject of analysis reports transmitted to the safety authorities. For the latter, since the 2010s, EDF deployed an in-depth event analysis method especially for identifying in-depth causes [COU 20]. We recall that high-risk industries encounter recurrent difficulties in identifying in-depth causes [CAI 03], and similarly in carrying out organizational event analyses [DIE 12a, LLO 18]. Finally, provision for learning the lessons of major accidents are deployed [DOE 05, IGS 20] or should be developed to better exploit the history and knowledge of accidents [DEC 10].

6.6.8. *Conducting organizational change*

Conducting organizational change has been the subject of recent preoccupation in the nuclear industry [DEC 11]. On the sociotechnical level, it extends the subject of the consideration of human and organizational impacts linked to technical modifications (see section 6.5.6). In 2003, the IAEA made the INSAG-18 report [IAE 03] available to the nuclear industry; the report warns of the risks induced by organizational change if conducted inappropriately. It is possible to cite an emblematic case of this type of deviation that happened at NASA that contributed, among other root causes to the accident of the space shuttle Columbia (2003). Indeed, *torrents of change* [CAI 03] destabilized management of the safety of shuttle flights. They were made in the spirit of continuous improvement [ROU 16], especially to "shake up" an agency perceived as "bureaucratic".

A little before 2010, the deployment of North American models of OCCs[14] (outage control centers) and maintenance reliability approach AP 913 [COU 20] benefitted from analyses using the SOH approach developed by EDF R&D [LEG 12], which scope was extended to organizational changes at the request of the safety authorities. These would be subject to inspection by the safety authority (ASN). The implementation of this SOH approach promoted the appropriation by agents of nuclear sites of the new mode of organizing the outage management. In addition, central departments with their advisory teams and peers from the operator's other sites were able to support sites in carrying out change. In the 2010s, the operator made many changes simultaneously (outage management organization, maintenance methods, management of contractors, making intervention practices reliable, harmonizing procedures, managing replacement parts and new information systems), in a context that is itself evolving (ageing of installations, renewal of personnel, search for competitiveness, increase in regulatory requirements) to which must be added actions carried out after Fukushima. Field inquiries by IRSN [COU 20] showed that although some changes were spread over time at each site to avoid cumulative effects on workload, the accumulated and joint effects of these changes on the organization of work could not always be anticipated and forestalled.

6.6.9. *Organizing maintenance and subcontractors' work*

Reactor outages are programmed to recharge the reactor core by renewing some of the fuel. The operator uses this period to carry out equipment maintenance, whether preventive or corrective. In France, around 80% of maintenance is carried out during these outages. Nevertheless, following North American practices, a transfer of some of these activities to the normal operation state has been underway for several years. Three types of outage are conducted: outages simply to recharge the fuel with light maintenance; outages for maintenance interventions and implementation of modifications; and 10-year intervals outages to test the containment pressures of the vessel and the reactor building. In these three situations, several thousand maintenance interventions are carried out by several hundred workers, the majority being contractors [COU 20].

The organizational complexity to manage outages is substantial and requires implementing an organizational infrastructure to prepare them and coordinate their realization in situations where unexpected events are common [DEC 14]. We recall that organizational complexity is a recurrent root cause of accidents [DIE 06, DEC 12]).

14 Called COPAT (Centre opérationnel de pilotage des arrêts de tranche) in France.

Until the 1990s, internal maintenance departments ensured this organization. From the mid-1990s, a project mode organization was adopted. Around 50 workers from operation and maintenance departments were thus made available for several months and were grouped on the same building floor. The outage project was managed by a team of a dozen people, gathered in a dedicated room and with modern communication tools equivalent to those in a crisis management room [COU 20].

Like foreign operators, EDF broke this outage preparation phase down into around 60 activities (or work packages) staggered over monthly stages. Among essential preparation capacities, one of the objectives was to control the volume and impact of late demands on workload that can destabilize planning and degrade the working conditions in which work is prepared and carried out. Capacities for regulation of request and workload management of outage project teams are some of the key provisions. The quality of technical exchanges between those preparing maintenance for different professional groups (electricity, mechanics, etc.), for operation and the project team during technical meetings to prepare each of the activities is essential and contractors should have contractual resources so that they can participate in the preparation meetings.

Piloting in project mode facilitates the work of coordinating and decision-making but adds interfaces, which complicates the organization. Shortcuts in the circulation of information and decision-making can weaken the structures put in place to control risks. The existence of relays in the field which are the "eyes of the project" then makes it possible to gain in anticipation time and reactivity. Anticipation relies especially on the development of planning, which mobilizes a specific team and many exchanges in view of daily updates of schedule following delays and slips in interventions [GLO 04, TIL 16, TIL 19, DEC 22].

In France, since the 1990s, EDF made the industrial choice to sub-contract up to 80% of interventions in some maintenance domains. Around 25,000 employees within more than 400 businesses take part in these interventions. Situations are very variable from one sub-contractor to another [PON 11]. Regulations prescribe that the operator should remain responsible for safety, but the safety authority enquires about the ability of the operator to remain responsible for safety in cases where this outsourcing causes a loss of skills. For a decade, the operator has "reinternalized" maintenance activities so as to have more skilled resources internally and to guarantee its status as the operator responsible for safety. Sub-contracted maintenance interventions have been subject to inspection since the 1990s as well as regular HOF evaluations (1990s, 2000s, 2010s) or discussion groups with the different stakeholders [COF 17a]. Many measures were taken and strengthened by EDF to manage the sub-contraction process [GAR 16]. Since the 2010s, EDF has ensured that businesses that respond to call to tenders have, in principle, the abilities required to intervene, because of audits of their management systems and qualifications conditioned for carrying out some work subject to monitoring. The contractual relationship is of an asymmetrical nature but to promote organizational learning of the respective parties, the logic of co-responsibility for success as well as failure should prevail. We note to finish that the conditions for coordination and contracting for a safe performance are the subject of research with many nuclear organizations, for example, with the RESOH research chair at Nantes [COU 17].

6.7. Cross-cutting aspects

6.7.1. *The challenges of integration, organization and time*

A number of cross-cutting dimensions to approaches aiming to make human and organizational actions more reliable can be highlighted. In the first place, studies carried out by HOF specialists are sometimes relatively separate from studies carried out by engineers, which results in a juxtaposition effect with regard to the sociotechnical system. Indeed, an organization should be adapted to operating needs for processes and devices, operational activities, skills and the environment, needs that evolve with time. Consequently, the challenge of integration remains entire. This objective is relatively easy to express. It particularly requires de-compartmentalizing expertise that is specialized, or indeed segmented by organizations. It is possible to consider that it is a facet and a risk linked to the bureaucratization of safety [DEK 14]. However, the integration of multiple expertises on a high-risk system is difficult to implement due to resources constraints on each area of professional expertise and to an excessively disciplinary management of transverse problems that appear in the design and use of nuclear reactors. This may also lead to separating analyses of technical factors, from human, organizational and social factors or to distinguishing micro-, meso- and macroscopic levels. Indeed, all dimensions are involved in the activity of actors in organizations. Similarly, management processes and provisions are designed and piloted in a segmented fashion. Indeed, during work, it is really all these provisions that sustain, cumulatively, the realization of activities. Work consists of knowing how to articulate in a singular manner, heterogeneous elements depending on the contingencies of the work situation. The challenge of integration in the activity remains and it should be the subject of more detailed descriptions, as needed, with new frameworks and theories of activity, work, knowledge, materiality and organizational processes [GHE 19, LOR 20].

The evolution of approaches leads to the wish to better understand the organizational processes that have transverse and multiplying effects on many work situations and activities in which risk management is constructed, deconstructed and is reconstructed daily. Although there has been a consensus for more than 20 years on the benefit of the "organizational paradigm" [REA 97, LLO 10b], various types of difficulty are encountered installing it. The term "organization" is polysemic (just like for "activity") and each discipline and each school of thought has a definition, concepts and experiences associated with it. This poses a problem when it involves, for example, analyzing the organization of risk management or "deep" or "root" causes, relating to organizational dysfunctions and today there are still significant gaps in representation [DIE 12, LLO 18]. In addition, organizational analysis addresses spatial dimension of an extended organizational network. Analysis should be deployed in a vertical and hierarchical perspective from operators to higher level deciders and their experts, and in a horizontal, transverse perspective, considering internal and external bodies [DIE 06]. However, the extent of the organizational networks of large groups or major industrial sites poses methodological challenges for understanding the organization's operation as a whole, especially through the prism of interactions between entities [LLO 18].

The time dimension is essential, whether to analyze the evolutions of work situations diachronically or to analyze organizations historically. Notable socio-historical analyses have been carried out during major industrial accidents [VAU 96, CAI 03, CSB 07] and more recently, after the Fukushima accident [DIE 12b] during research on the processes of regulating flood risks [MAN 18] and earthquake risks [ROG 20] or the risks of severe accident [GOU 21] in the context of the AGORAS research project. The proactive search for latent deficiencies, gradual deviations by organizations, weak signals, in relation to the potential existence of an incubation period is an approach that should be developed. Similarly, anticipation of future risks should allow space for new approaches to open frameworks of thought and imagination as well as temporalities over the period of several years [ESR 20]. Analyses of "organizing" processes, especially "sensemaking" processes [WEI 95] are a way of grasping the dynamic of interactions, information flows and the creation of sense in line with decision-making processes, and activities as something continuous becoming.

6.7.2. *The contribution of the systemic approach*

The challenge of integrating different dimensions in interrelationship and interactions in extended organizational networks while considering temporality, leads to underlining the benefit of the systemic view of risk management approaches. The IAEA has recently recognized the benefit of developing this type of view: a systemic approach is defined by the IAEA [IAE 16a] as an approach envisaging the system as a whole in which interactions between technical, human and organizational factors are duly considered. Many concepts of HOF approaches form part of the current of systemic thought, but not all HOF specialists are necessarily aware of this filiation [VAU 18b].

The concept of interaction can express flows of materials, energy or information or even relationships between elements [VAU 19b]. The flows are represented in descriptive models of the functioning of work systems, relationships of adaptation and causality, in particular, are represented in explanatory models of this functioning [VAU 18c]. As illustrations, we cite representations of man–machine systems that present the flow of actions of an operator on a machine and the flow of information that they use to make decisions and then carry out actions [SEN 95]. In relation to adaptation and causality, these are at the very heart of ergonomic thought. From 1970 to 1988, the definition of ergonomics proposed by the French language ergonomics society[15] explicitly used these notions:

> *ergonomics can be defined as the adaptation of work to the human being, more precisely, as the implementation of scientific knowledge on humans necessary to design tools, machines and devices that can be used with maximum comfort, safety and efficiency.*

Finally, these relationships of adaptation and causality are found in particular in models [LEP 01] [GUE 97] or indeed in the CEA micro-macro model inside and between a work situation and organizational factors [TOS 12, VAU 18c]. The concept of interaction can serve as a support for different approaches such as system dynamics or known as structured analyses [VAU 19b], for example, the STAMP (Systems-Theoretic Accident Model and Processes) method proposed by Nancy Leveson [LEV 11, HAR 12] or decision analysis of complex systems [BUC 02]. In structured approaches, interaction is formed from two types of exchanges: piloting actions by a piloted actor, carried out by a "pilot" actor; operation feedback on the activity of the piloted actor, recovered by the pilot actor. The pilot actor "pushes/generates" actions and "requests" operation feedback from the piloted actor. In the context of this type of modeling, each pilot actor can itself be linked to an actor that pilots it. Similarly, a piloted actor can also be an actor who pilots other actors. This concept of interaction can also be expressed through the notion of a dialogical pair or an ago-antagonist one [MOR 96, BER 02]. Two elements of this type of pair are both opposed and complementary.

In addition to the concept of interaction, another essential concept is to consider the system effects as whole, which is found implicitly in different models already mentioned [LEP 01, REA 06]. These models relate especially to the influence of all the factors or components of a work situation on the realization of the operators' actions. These models are also interested in the structure and operating rules of the organization, which then leads in this case to considering a "whole" still broader than the previous one [VAU 18b]. In this context, the concept of "requisite variety" by Ashby [ASH 56] is essential since he encourages us to have a diversity of views

15 Société d'ergonomie de langue française (SELF).

on a system to govern it. The concept of the "whole" can thus be expressed too in the context of the framework of a global, temporal (or diachronic) vision. For example, the evolution of organizations would pass through different stages, interspersed with crises: creativity, direction, delegation, co-ordination, collaboration and extra-organizational solutions [GRE 98]

6.7.3. *Reflexivity and critical approach*

The level of reflexivity of HOF approaches has increased in view of the experience acquired, benefits but also some failures. An additional level of maturity is worth to be acquired by developing a more critical look at our methods, our approaches, our risk management and governance arrangements. This requires spaces to be established to discuss the work [ROC 16], feedback and sharing of experience [CAS 16] and nuclear safety [BRI 20]. This requires the organization of debates, involving peers internally and externally with professionals in risk management[16], potential whistleblowers, external stakeholders in fields that are controversial by nature, to re-question some risk management hypotheses, but also cultural and regulatory effects that can result in excess and a lack of attention to some risks. Risks, management and governance of risks are in fact socially constructed [SHO 84] and they are the subject of technical discussions and scientific disputes. Finally, a mindset that leads itself to enquire to remove doubts and improve safety should be encouraged. New risks present themselves; attention and vigilance should be maintained by mechanisms to be developed or reinforced.

6.7.4. *HOF specialists and HOF relays: the contribution of HOF networks*

HOF specialists have assumed responsibility for deploying HOF approaches in the nuclear industry [ROL 09]. To multiply this deployment and this appropriation in day-to-day practices, designers, operators, experts and regulators, HOF networks have been established since the middle of the 2000s by operators and regulators. EDF some years ago increased the number of HOF specialists on nuclear sites by providing one HOF specialist for a pair of nuclear reactors. HOF relays have been trained in reliability practices, especially to implement them in activities. The HOF network at CEA includes HOF specialists at local and central levels, as well as HOF relays in nuclear installations. These latter have fewer skills than previous actors, but

16 We can mention experts with professional associations (in France, such as the IMdR, the ICSI, the COFSOH, and at European level, the ESReDA, ESRA, SRA, etc.) and more largely in engineering and human and social sciences.

can ensure implementation and monitoring at their installation, results of HOF studies carried out by specialists [VAU 19a].

6.8. Conclusion and perspectives

HOF have been considered for 40 years in the nuclear field, but has this disciplinary field of nuclear reliability and safety reached its maturity? The applied discipline of HOFs, or a collection of methodological practices developed by specialists from the human and social sciences, is structured around the analysis of human performance and factors that have an influence on this. These factors are found within work situations as well as in organizations' structure, interactions and operating modes. We note that social, organizational and human dimensions also influence the engineering of technical devices. The influence of some accidents in the 1970s and especially the Three Mile Island accident made it possible to initiate the development of knowledge on work situations (procedures, skills, man–machines interfaces) and organizations (operation feedback, design process, organization of teams, crisis management). The accidents at Chernobyl and Fukushima have given impulses to themes such as safety management and culture, risk management and resilience.

With the decisive support of research integrated within the nuclear industry and cooperation with academic research, frameworks and reliability methods for work situations and organizations have also been developed. They have contributed to the improvement of organizations' functioning and to consolidating good practices developing there [COU 17]. We have outlined some means of progress (e.g. the integration of different dimensions in activity, systemic, organizational aspects, HOF networks, and so on) but many challenges remain: for the production of methods and knowledge on work situations and organizations (non-technical skills; to evaluate the effectiveness of training or measures taken following HOF studies, etc.), the refinement of new methods for HOF analysis, or the development of a professional sector for HOF specialists. Orientations and priorities for developing particular knowledge could be debated. HOFs have essentially been influenced by the ergonomics, psychology and sociology of organizations. Borrowings from other disciplinary fields could be strengthened (anthropology, management sciences, sociology of technology and history).

Nevertheless, studies carried out by HOF specialists remain in the minority with regard to design and operations engineering studies. Although there have been notable improvements from one decade to another, evolution remains slow, and it is an evolutionary rather than a revolutionary type. Indeed, some frameworks, concepts or questions should lead to more significant changes in professional practices. Accidents and regulatory changes drive change and strengthening of requirements, but lessons from accidents are sometimes missed and forgotten. On some points, it is possible to observe the risk of regression. Nothing is therefore gained. Moreover, one of the challenges consists of marking the specifics of HOF approaches with regard to areas of concern, such as quality, operational excellence and psycho-social risks. The overall objectives are close and some of the methods used are sometimes similar. The risk could then be a dilution of HOF approaches in these areas of concern.

The nuclear industry and society are undergoing evolutions that lead to new questions and present opportunities but also risks for risk management: the ageing of installations and equipment, generational changes, the transition to digital, climate change, the impact of a pandemic crisis such as Covid-19 on work organizations and risk management, malicious and terrorist acts, post-accident situations and waste storage over very long durations. Consideration of human and organizational dimensions is not acquired during the design stage due to the rhythm of changes and it can then happen that HOF specialists are only requested when problems arise. The consolidation of knowledge and practices must also be articulated with exploration of new difficulties.

6.9. References

[ABR 99] ABRAMOVICI M., La prise en compte de l'organisation dans l'analyse des risques industriels : méthodes et pratiques, PhD Thesis, ENS, Cachan, France, 1999.

[ADA 06] ADAMS D., A Layman's introduction to human factors in aircraft accident and incident investigation, Australian Transportation Safety Bureau – ATSB Safety information paper B2006/0094, 2006.

[ALE 18] ALENGRY J., FALZON P., DE LA GARZA C. *et al.*, "What is 'training to cope with crisis situations'? Developing a reflexive training device for a crisis support team", *Proceedings of the 20th Congress of the International Ergonomics Association, Florence, Italy, August 26-30, 2018, Volume V: Human Simulation and Virtual Environments, Work with Computing Systems (WWCS), Process Control*, Springer Internanational Verlag, Cham, Switzerland, pp. 60–67, 2018.

[AMA 13] AMALBERTI R., *Navigating Safety, Necessary Compromises and Trade-offs– Theory and Practice*, Springer, The Netherlands, 2013.

[ASH 56] ASHBY W.R., An introduction to cybernetics, available at: http://pespmc1.vub.ac.be/books/IntroCyb.pdf, 1956.

[ASN 12] ASN, Décision no. 2012-DC 0274 à 0292 du 26 juin 2012 suite aux ECS, 2012.

[ASN 17] ASN, Guide n°22 relatif à la "Conception des réacteurs à eau sous pression", 18 July 2017.

[BER 02] BERNARD-WEIL E., *Approche des systèmes ago-antagonistes, AG 1575*, Techniques de l'Ingénieur, Paris, France, 2002.

[BER 14] BERNARD B., *Comprendre les facteurs humains et organisationnels – Sûreté nucléaire et organisations à risques*, EDP Sciences, Les Ulis, France, 2014.

[BES 17] BESNARD D., BOISSIERES I., DANIELLOU F. *et al.*, La culture de sécurité : comprendre pour agir, coordination, *Cahiers de la sécurité industrielle*, no. 2017-01, ICSI, 2017.

[BIE 13] BIEDER C., BOURRIER M. (eds), *Trapping Safety into Rules – How Desirable or Avoidable is Proceduralization?* Ashgate Publishing, Aldershot, UK, 2013.

[BLA 16] BLATTER C., DECHY N., GARANDEL S., Vers un retour d'expérience prenant en compte les facteurs organisationnels et humains, IMdR, available at: https://www.imdr.eu, 2016.

[BÖH 98] BÖHLE F., MILKAU B., *De la manivelle à l'écran – L'évolution de l'expérience sensible des ouvriers lors des changements technologiques*, Éditions Eyrolles, Paris, France, 1998.

[BOU 99] BOURRIER M., *Le nucléaire à l'épreuve de l'organisation*, PUF, Paris, France, 1999.

[BOU 01] BOURRIER M., *Organiser la fiabilité*, Éditions L'Harmattan, Paris, France, 2001.

[BRI 16] BRINGAUD V., JOURNÉ B., MBAYE S. *et al.* (eds), *Le retour d'expérience dans les organisations à risques*, Presses des MINES, Paris, France, 2016.

[BRI 20] BRINGAUD V., GUILLAUME O., LOT N., Espaces de discussion, apprentissages et sûreté nucléaire, *Education Permanente* n°224/2020-3, 2020.

[BUC 02] BUCKI J., *Analyse décisionnelle des systèmes complexes*, Techniques de l'Ingénieur, Saint-Denis, France, 2002.

[CAI 03] CAIB, Report Volume 1 of Columbia Accident Investigation Board, National Aeronautics and Space Administration, Washington DC, available at: http://caib.nasa.gov, 2003.

[CAS 16] CASSE C., CAROLY S., "Un instrument de construction de la sécurité", in Bringaud V., Journé B., Mbaye S. *et al.*, *Le Retour d'Expérience dans les organisations à risques*, Presses des MINES, Paris, France, 2016.

[CLA 99] CLARKE L., *Mission Improbable, Using Fantasy Documents to Tame Disasters*, University of Chicago Press, Chicago, USA, 1999.

[COF 17a] COFSOH, Pour une contribution positive de la maintenance sous-traitée à la sûreté nucléaire, available at: https://www.asn.fr/L-ASN/Comite-sur-les-facteurs-sociaux-organisationnels-et-humains, 2017.

[COF 17b] COFSOH, COFSOH Pour une approche intégratrice des facteurs organisationnels et humains (FOH) lors de la gestion d'une crise dans le nucléaire, available at: https://www.asn.fr/L-ASN/Comite-sur-les-facteurs-sociaux-organisationnels-et-humains, 2017.

[COF 19] COFSOH, Développer la sécurité, available at: https://www.asn.fr/L-ASN/Comite-sur-les-facteurs-sociaux-organisationnels-et-humains, 2019.

[COU 17] COUTURIER J., SCHWARZ M. (eds), *État des recherches dans le domaine de la sûreté des réacteurs à eau sous pression*, EDP Sciences, Paris, France, available at: https://www.irsn.fr/FR/Larecherche/Actualites_Agenda/Actualites/Documents/RDreacteurs_WEB_NetB.pdf, 2017.

[COU 20] COUTURIER, J. (ed.), *Éléments de sûreté nucléaire, les réacteurs à eau sous pression, sous la direction et coordination*, EDP Sciences, Paris, France, available at: https://www.irsn.fr/FR/Larecherche/publications-documentation/collection-ouvrages-IRSN/Documents/livre-libman_PDF%20web%20couleur.pdf, 2020.

[CRU 83] CRU D., DEJOURS C., "Les savoir-faire de prudence dans les métiers du bâtiment, nouvelle contribution de la psychopathologie du travail à l'analyse des accidents et de la prévention dans le bâtiment", *Les Cahiers médico-sociaux*, no. 3, pp. 239–247, 1983.

[CSB 07] CSB, Investigation report, refinery explosion and fire, BP – Texas City, Texas, USA, March 23, 2005, Report No. 2005-04-I-TX, 2007.

[DAN 10] DANIELLOU F., SIMARD M., BOISSIERES I., Les facteurs humains et organisationnels de la sécurité industrielle : un état de l'art, FonCSI, *Les cahiers de la sécurité industrielle*, Toulouse, France, 2010.

[DEC 08] DECHY N., DIEN Y., LLORY M., Les échecs organisationnels du retour d'expérience, INERIS report, No. DRA-08-95321-15660A, 2008.

[DEC 10] DECHY N., DIEN Y., LLORY M., "Pour une culture des accidents au service de la sécurité industrielle", *Congrès IMdR $\lambda\mu 17$*, 5–7 October, La Rochelle, France, 2010.

[DEC 11] DECHY N., ROUSSEAU J.-M., LLORY M., "Managing sociotechnical changes: hard lessons learned from accidents and further challenges", *Proceedings of the 40th ESReDA Seminar*, 25–26 May, Bordeaux, France, 2011.

[DEC 12] DECHY N., DIEN Y., LLORY M., "La complexité des systèmes sociotechniques à risques rend-t-elle les accidents inévitables ?" *Congrès IMdR $\lambda\mu 18$*, 16–18 October, Tours, France, 2012.

[DEC 14] DECHY N., THELLIER S., ROUSSEAU J.-M. *et al.*, "Gestion des aléas lors des activités de maintenance : maîtriser les risques malgré l'incertitude", *Congrès IMDR $\lambda\mu 19$*, 21–23 October, Dijon, France, 2014.

[DEC 16] DECHY N., DIEN Y., MARSDEN E. et al., "Un retour d'expérience dynamique à l'épreuve de la temporalité", *Congrès IMdR $\lambda\mu 20$*, 11–13 October, Saint-Malo, France, 2016.

[DEC 18] DECHY N., BLATTER C., "Mieux prendre en compte les facteurs organisationnels et humains dans le retour d'expérience : quels apports et défis pour le big data et le traitement automatique du langage", *Congrès IMdR $\lambda\mu 21$*, 16–18 October, Reims, France, 2018.

[DEC 22] DECHY N., LARGIER A., "Workload planning management of maintenance activities in nuclear power plants: compensation mechanisms at the contractor interface", in: Hayes J., Tillement S. (eds), *Contracting and Safety: Exploring Outsourcing Practices in High-hazard Industries*, Springer, Cham, Switzerland, 2022.

[DEJ 95] DEJOURS C., *Le facteur humain. Que sais-je ?* Editions PUF, Paris, France, 1995.

[DEK 06] DEKKER S., *The Field Guide to Understanding Human Error*, Ashgate Publishing, Aldershot, UK, 2006.

[DEK 07] DEKKER S., *Just Culture: Balancing Safety and Accountability*, Ashgate Publishing, Aldershot, UK, 2007.

[DEK 14] DEKKER S., "The bureaucratization of safety", *Safety Science Journal*, vol. 70, pp. 348–357, 2014.

[DEL 13] DELEUZE G., IPPERTI P., *L'analyse des risques. Concepts, outils, gestion, maîtrise*, Editions EMS Management Société, Caen, France, 2013.

[DIE 98] DIEN Y., "Safety and application of procedures, or how do they have to use operating procedures in nuclear power plants?" *Safety Science*, vol. 29, pp. 179–187, 1998.

[DIE 06] DIEN Y., "Les facteurs organisationnels des accidents industriels", in: Magne L., Vasseur D. (eds), *Risques industriels – Complexité, incertitude et décision : une approche interdisciplinaire*, Éditions TEC & DOC, Lavoisier, Paris, France, 2006.

[DIE 12a] DIEN Y., DECHY N., GUILLAUME E., "Accident investigation: from searching direct causes to finding in-depth causes. Problem of analysis or/and of analyst?", *Safety Science*, vol. 50, no. 6, pp. 1398–1407, 2012.

[DIE 12b] DIET, The national diet of Japan, The Fukushima Nuclear Accident Independent Investigation Commission report, available at: https://warp.da.ndl.go.jp/info:ndljp/pid/3856371/naiic.go.jp/en/report/, 2012.

[DOE 05] DOE (Department Of Energy), Action-plan – Lessons learned from the Columbia space shuttle accident and Davis-Besse reactor pressure-vessel head corrosion event, available at: https://www.dnfsb.gov/sites/default/files/document/1221/ltr_2005729_6821.pdf, 2005.

[ECH 18] ECHAJARI L., Apprentissage organisationnel à partir d'expériences rares et complexes : le rôle de la codification des connaissances. Le cas de deux accidents nucléaires, PhD Thesis, Université Côté d'Azur, France, 2018.

[EDF 06] EDF, Les pratiques de l'intervenant – Performance Humaine, Report, EDF, March 2006.

[EDF 16] EDF, Mémento sûreté nucléaire en exploitation, EDF, 2016.

[EDM 99] EDMONDSON A.C., "Psychological safety and learning behavior in work teams", *Administrative Science Quarterly*, vol. 44, no. 2, pp. 350–383, 1999.

[ELB 05] ELBARADEI M., "The enduring lessons of Chernobyl", *Proceedings of the IAEA International Conference "Chernobyl: Looking Back to Go Forward"*, 6–7 September, 2005.

[ESR 20] ESREDA, ESREDA enhancing safety: the challenge of foresight, ESReDA Project Group Foresight in Safety (ed.), available at: www.esreda.org, 2020.

[EYD 17] EYDIEUX J., Gouverner les risques par le doute. Une approche pragmatiste du dialogue technique, Thesis, Université de Nantes, Nantes, France, 2017.

[FAS 11] FASSERT C., FAYE H., "Que faire de la culture de sûreté ?", *46ème Congrès de la SELF*, 14–16 September, Issy-les-Moulineaux, France, 2011.

[FAV 70], FAVERGE, J., "L'homme, agent d'infiabilité et de fiabilité du processus industriel", *Ergonomics*, vol. 13, no. 3, 1970.

[FLA 19] FLANDIN, S., POIZAT G., PERINET R., "Contribuer à la sécurité industrielle 'par le facteur humain' – Un regard pour aider à (re)penser la formation", *FonCSI*, no. 2019-01, 2019.

[FOR 16] FORNETTE M.-P., JOLLANS J.-Y., *Former les équipes à la sécurité et à la performance avec le Crew Resource Management*, Editions Octares, Toulouse, France, 2016.

[FUC 12] FUCKS I., "L'énigme de la culture de sécurité dans les organisations à risques : une approche anthropologique", *Le Travail Humain*, vol. 75, pp. 399–420, 2012.

[FUC 13] FUCKS I., DIEN Y., "'No rule no use'? The effects of over-proceduralization", in: Bieder C., Bourrier M. (eds), *Trapping Safety into Rules – How Desirable or Avoidable is Proceduralization?*, Ashgate Publishing, Aldershot, UK, 2013.

[GAR 14] GARCIAS F., Apprentissage, désapprentissage et réapprentissage organisationnels : le cas d'une activité d'ingénierie de grands projets complexes, Thesis, Ecole Nationale Supérieure des Mines de Paris, France, 2014ENMP0090, 2014.

[GAR 16] GARRON J., "Improving control over subcontracted operations at nuclear power plants in France", *EUROSAFE Forum*, 7–8 November, Munich, Germany, 2016.

[GHE 19] GHERARDI S., *How to Conduct a Practice-based Study: Problems and Methods*, 2nd ed., Edward Elgar Publishing Ltd, Cheltenham, UK, 2019.

[GIS 15] GISQUET E., OLDER M., A human and organizational factors perspective on the Fukushima nuclear accident, Report IRSN-PSN-SRDS-SFOHREX-2015-01, 2015.

[GIS 16] GISQUET E., LEVY E., JEFFROY F., Appréhender les aspects culturels des organisations dans les industries à risques, Report IRSN PSN-SRDS/SFOHREX n°2016-002-1, 2016.

[GLO 04] GLOBOKAR T., "Compromis temporels dans la gestion des projets. Le cas de la maintenance nucléaire", *Revue Française de Gestion*, vol. 152, pp. 81–96, 2004.

[GOF 87] GOFFMAN E., *Façons de parler*, Editions de Minuit Paris, France, 1987.

[GOU 21] GOUMRI M., Apprivoiser le cygne noir. Construction et circulation des savoirs et des ignorances dans le gouvernement de l'accident nucléaire majeur. PhD Thesis, Université de Paris, France, 2001.

[GRE 98] GREINER L.E., Evolution and revolution as organisations grow, *Harvard Business Review*, pp. 55–67, vol. 76, no. 3, May–June, 1998.

[GUE 97] GUERIN F., LAVILLE A., DANIELLOU F. *et al.*, *Comprendre le travail pour le transformer*, Éditions ANACT, Lyon, France, 1997.

[GUI 11] GUILLAUME E., Identifying and responding to weak signals to improve learning from experience in high-risk industries, PhD Thesis, TU Delft, The Netherlands, 2011.

[GUL 00] GULDENMUND F.W., "The nature of safety culture: a review of theory and research", *Safety Science*, vol. 34, nos 1–3, pp. 215–257, 2000.

[HAR 12] HARDY K., GUARNIERI F., *Modéliser les accidents et les catastrophes industrielles : la méthode STAMP*, SRD, Éditions Lavoisier, Cachan, France, 2012.

[HOL 14] HOLLNAGEL E., *Safety-I and Safety-II: The Past and Future of Safety Management*, Ashgate Publishing, Hampshire, UK, 2014.

[HOL 06] HOLLNAGEL E., WOODS D., LEVESON N.G., *Resilience Engineering: Concepts and Precepts*, Ashgate Publishing, Aldershot, UK, 2006.

[HOP 01] HOPKINS A., "Was three mile island a normal accident?", *Journal of Contingencies and Crisis Management*, vol. 9, no. 2, pp. 65–72, 2001.

[HOP 03] HOPKINS A., *Lessons from Longford. The Esso Gas Plant Explosion*, 7th ed., CCH Australia Limited, North Ryde, Austrailia, 2003.

[HOU 02] HOURTOULOU D., Projet ASSURANCE – Assessment of the uncertainties in risk analysis of chemical establishments, E. C. project ENV4-CT97-0627, Report N° DRA-07 – INERIS – Dho-2002-26824, 2002.

[IAE 91] IAEA, Safety culture – Series No.75-INSAG4, Report, International Nuclear Safety Advisory Group, Vienna, Austria, 1991.

[IAE 03] IAEA, Managing change in the nuclear industry: the effects on safety – INSAG18, Report, by the International Nuclear Safety Advisory Group, Vienna, Austria, 2003.

[IAE 13] IAEA, Report on human and organizational factors in nuclear safety in the light of the accident at the Fukushima Daiichi Nuclear Power Plant, Report, International Experts Meeting, 21–24 May, Vienna, Austria, 2013.

[IAE 16a] IAEA, Leadership and management for safety, General Safety Requirements/No. GSR Part 2, IAEA Safety Standards, 2016.

[IAE 16b] IAEA, *Proceedings of the International Conference on Human and Organizational Aspects of Assuring Nuclear Safety – Exploring 30 Years of Safety Culture*, 22–26 February, Vienna, Austria, 2016.

[IGS 21] IGSNR, Rapport de l'Inspecteur Général pour la Sûreté Nucléaire et la Radioprotection, Report, EDF, Année 2020, 2021.

[IMD 13a] IMDR, Méthodes d'analyse textuelle pour l'interprétation des REX humains, organisationnels et technique, IMdR-P10-5, available at: www.imdr.fr, 2013.

[IMD 13b] IMDR, Détection et pertinence d'un signal faible dans le traitement d'un retour d'expérience, IMdR-P12-1, available at: www.imdr.fr, 2013.

[IRS 18] IRSN, Avis IRSN/2018-00178 du 29/06/2018, Examen des dispositions de formation et de préparation aux situations extrêmes stressantes mises en œuvre par EDF – Le cas de la FARN, 2018.

[IRS 21] IRSN, Anticipation et résilience : réflexions dix ans après l'accident de Fukushima Daiichi, Report IRSN 2021-00001, 2021.

[JOU 99] JOURNÉ B., Les organisations complexes à risques : gérer la sûreté par les ressources. Études de situations de conduite de centrales nucléaires, PhD Management Thesis, École Polytechnique, Palaiseau, France, 1999.

[JOU 05] JOURNÉ B., "Étudier le management de l'imprévu : méthode dynamique de l'observation in situ", *Finance Contrôle Stratégie*, vol. 8, no. 4, pp. 63–91, 2005.

[JOU 06] JOURNÉ B., "Observer la sûreté et en rendre compte : éléments de méthode", in: Jeffroy F., Theureau J., Haradji Y. (eds), *Relation entre activité individuelle et activité collective : confrontation de différentes démarches d'études*, Editions Octarès, Toulouse, France, 2006.

[JOU 14] JOUNIAUX P., HADIDA D., DECHY N. *et al.*, "Détection, pertinence et amplification des signaux faibles dans le traitement du retour d'expérience", *Congrès IMdR λμ19*, 21–23 October, Dijon, France, 2014.

[KEM 79] KEMENY J.G., BABBITT B., HAGGERTY P.E. *et al.*, The need for change – The legacy of TMI, Report of the President's Commission on The Accident At Three-Mile Island, Government Printing Office, Washington, DC, USA, 1979.

[KOU 00] KOUABENAN D.R., *Explication naïve de l'accident et prévention*, PUF, Paris, France, 2000.

[LAG 91] LAGADEC P., *La gestion des crises – Outils de réflexion à l'usage des décideurs*, McGraw-Hill, New York City, USA, 1991.

[LAG 11] LAGRANGE V., "Culture de sûreté, concept fourre-tout ou opportunité pour tenir compte davantage des Hommes et des Organisations dans les industries à risques", *46ème Congrès de la SELF*, 14–16 September, Issy-les-Moulineaux, France, 2011.

[LAP 82] LAPORTE T., "On the design and management of nearly error-free organizational control systems", in: Sills D.L., Wolf C.P., Shelanski V.B. (eds), *A Westview Special Study: Accident at Three Mile Island*, The Human Dimensions, Boulder, Colorado, USA, 1982.

[LAR 08] LARGIER A., DELGOULET C., DE LA GARZA C., Quelle prise en compte des compétences collectives et distribuées dans la gestion des compétences professionnelles ? *PISTES 10–1*, 2008.

[LAV 91] LAVE J., WENGER E., *Situated Learning: Legitimate Peripheral Participation*, Cambridge University Press, Cambridge, UK, 1991.

[LEG 98] LE GUILCHER B., DIEN Y., "Palier N4 : apports de l'ergonomie dans la conception de la salle de commande", *Revue générale nucléaire*, vol. 1, pp. 27–32, 1998.

[LEG 09] LEGER A., DUVAL C., WEBER, P., "Methodological developments for probabilistic risk analyses of socio-technical systems", *Journal of Risk and Reliability*, vol. 223, no. 4, pp. 313–332, 2009.

[LEG 12] LE GUILCHER B., LAGRANGE V., La démarche SOH – Ambition, Avancement, Journées Techniques de la Société Française d'Energie Nucléaire (SFEN), Paris, France, 2012.

[LEM 77] LE MOIGNE J.-L., *La théorie du système général, théorie de la modélisation* (2nd edition in 1994), PUF, Paris, France, 1977.

[LEP 01] LEPLAT J., *L'analyse du travail en psychologie ergonomique (Tome I)*, 2nd ed., Octarès Éditions, Toulouse, 2001.

[LEV 11] LEVESON N.G., "Applying systems thinking to analyze and learn from events", vol. 49, no. 1, pp. 55–64, 2011.

[LES 01] LESCA H., "Veille stratégique : passage de la notion de signal faible à la notion de signe d'alerte précoce", *Proceedings of the VSST Conference*, Barcelona, Spain, 2001.

[LLO 96] LLORY M., *Accidents industriels : le coût du silence – Opérateurs privés de parole et cadres introuvables*, Éditions L'Harmattan, Paris, France, 1996.

[LLO 99] LLORY M., *L'accident de la centrale nucléaire de Three Mile Island*, Éditions L'Harmattan, Paris, France, 1999.

[LLO 06] LLORY M., DIEN Y., Les systèmes sociotechniques à risques : une nécessaire distinction entre fiabilité et sécurité, *Performances*, no. 30, 31, 32, 2006.

[LLO 10a] LLORY M., DIEN Y., *Analyse organisationnelle de la sécurité*, Technique de l'Ingénieur, AG 1577, 2010.

[LLO 10b] LLORY M., MONTMAYEUL R., *L'accident et l'organisation*, Éditions Préventique, Bordeaux, France, 2010.

[LLO 18] LLORY M., MONTMAYEUL R., *Comprendre l'accident – La pratique de l'analyse organisationnelle de la sécurité*, Éditions L'Harmattan, Paris, France, 2018.

[LOR 20] LORINO P., *Pragmatisme et étude des organisations*, Editions Economica, Paris, France, 2020.

[MAG 13] MAGUIRE S., HARDY C., "Organizing processes and the construction of risk: a discursive approach", *Academy of Management Journal*, vol. 56, no. 1, pp. 231–255, 2013.

[MAN 18] MANGEON M., Conception et évolution du régime français de régulation de la sûreté nucléaire (1945–2017) à la lumière de ses instruments : une approche par le travail de régulation, PhD Thesis, Mines ParisTech, Paris, France, 2018.

[MAN 20] MANGEON M., DECHY N., ROUSSEAU J.-M., Les accidents nucléaires de 1969 et 1980 à Saint-Laurent-des-Eaux : quand la transition engendre l'oubli, *eCongrès IMdR λμ22*, 2020.

[MAR 14] MARSDEN E. (ed.), Quelques bonnes questions à se poser sur son dispositif de retour d'expérience, *Les cahiers de sécurité industrielle*, 2014-01, FonCSI, Toulouse, France, available at: www.foncsi.org, 2014.

[MAR 15] MARSDEN E., DECHY N., DIEN Y. *et al.*, Barriers to learning from incidents and accidents, ESReDA, available at: www.esreda.org, 2015.

[MAT 02] MATAHRI N., BAUMONT G., HOLBE C., Modèle d'analyse des incidents RECUPERARE, intégrant les facteurs techniques, humains et organisationnels, IRSN, Scientific and Technical Report, 2002.

[MAZ 93] MAZEAU, M., "L'homme, agent de fiabilité faillible", *Performances Humaines & Techniques*, vol. 66, pp. 24–29, 1993.

[MBA 09] MBAYE S., KOUABENAN D.R, SARNIN P., L'explication naïve et la perception des risques comme des voies pour améliorer les pratiques de REX : des études dans l'industrie chimique et l'industrie nucléaire, *Cahiers de la sécurité industrielle*, vol. 8, Toulouse, France, 2009.

[MER 10] MERAD M., *Aide à la décision et expertise en gestion des risques*, Éditions Lavoisier, Cachan, France, 2010.

[MER 16] MERAD M., DECHY N., DEHOUCK L. *et al.*, "Les décisions face aux risques majeurs, retours d'expérience et pistes d'améliorations", in: Merad M., Dechy N., Dehouck L. *et al.*, *Risques majeurs, incertitudes et décisions*, MA-Éditions – ESKA, Paris, France, 2016.

[MIN 89] MINTZBERG H., *Le management – Voyage au centre des organisations*, Édition d'Organisation, Paris, France, 1989.

[MON 06] MONTMAYEUL R., Les pressions de production : l'équilibre production-sécurité, Séminaire de Saint-André "Risques industriels et sécurité : les organisations en question", Llory M. (ed.), 26–27 September, Bompas, France, 2006.

[MON 13] MONTMAYEUL R., *1er Juin 1974. L'accident de l'usine chimique de Flixborough, Royaume-Uni*, Éditions Préventique, Bordeaux, France, 2013.

[MOR 77] MORIN E., *La méthode, la nature de la nature*, Éditions du Seuil, Paris, France, 1977.

[MOR 96] MORIN E., *Pour une réforme de la pensée*, Éditions Nathan, Paris, France, 1996.

[OCD 09] OCDE, The role of human and organizational factors in nuclear power plant modifications, CSNI Technical Opinion Papers, no. 10, Éditions OCDE, Paris, France, available at: https://doi.org/10.1787/d5e16496-en, 2009.

[OCH 78] OCHANINE D.A., "Le rôle des images opératives dans la régulation des activités de travail", *Psychologie et éducation*, vol. 2, pp. 63–54, 1978.

[PER 82] PERROW C., The president's commission and the normal accident, in: SILLS D.L., WOLF C.P., SHELANSKI V.B. (eds), *A Westview Special Study: Accident at Three Mile Island: The Human Dimensions*, Boulder, Colorado, USA, 1982.

[PER 84] PERROW C., *Normal Accidents. Living with High-Risk Technology*, Basic Books, New York, 1984, 2nd ed., Princeton University Press, Princeton, 1999.

[PON 11] PONNET M., Les relations de sous-traitance et leurs effets sur la sûreté et la sécurité dans deux entreprises : SNCF et GrDF, PhD Thesis, Université de Nantes, France, 2011.

[POW 16] POWER M., Riskwork: the organizational life of risk management. Introduction, in: Power M. (ed.), *Riskwork: Essays on the Organizational Life of Risk Management*, Oxford University Press, Oxford, UK, 2016.

[RAS 83] RASMUSSEN J., "Skills, rules, and knowledge – signals, signs, and symbols, and other distinctions in human performance models", *IEEE Transactions on Systems, Man, and Cybernetics*, vol. 13, no. 3, pp. 257–266, 1983.

[RAS 97] RASMUSSEN J., "Risk management in a dynamic society: a modelling problem", *Safety Science*, vol. 27, nos 2–3, pp. 183–213, 1997.

[REA 90] REASON J., *Human Error*, Cambridge University Press, Cambridge, UK, 1990.

[REA 97] REASON J., *Managing the Risks of Organizational Accidents*, Ashgate Publishing, Aldershot, UK, 1997.

[REA 06] REASON J., HOLLNAGEL E., PARIES J., Revisiting the "Swiss Cheese" model of accidents, Eurocontrol, available at: https://www.eurocontrol.int/publication/revisiting-swiss-cheese-model-accidents, 2006.

[ROB 93] ROBERTS K., *New Challenges to Understanding Organizations*, Macmillan Publishing Company, New York, USA, 1993.

[ROC 16] ROCHA R., MOLLO V., DANIELLOU F., Les espaces de débat sur l'activité réelle, in: Bringaud V., Journé B., Mbaye S. *et al.* (eds), *Le Retour d'Expérience dans les organisations à risques*, Presses des MINES, Paris, France, 2016.

[ROG 80] ROGOVIN M., FRAMPTON G., *Three-Mile Island: A Report to the Commissioners and to the Public*, Vol. 1, Nuclear Regulatory Commission, available at: https://www.hsdl.org/?abstract&did=778476, US NRC, Washington, USA, 1980.

[ROG 20] ROGER M., Construire la robustesse face aux risques imprédictibles : le cas du risque sismique pour la sûreté des centrales nucléaires françaises, PhD Thesis, Université Paris-Descartes, Paris, France, 2020.

[ROL 09] ROLINA G., *Sûreté nucléaire et facteurs humains – La fabrique française de l'expertise*, Presse des Mines, Paris, France, 2009.

[ROU 03] ROUX-DUFORT C., *Gérer et décider en situation de crise : outils de diagnostic, de prévention et de décision*, Dunod, Paris, France, 2003.

[ROU 08a] ROUSSEAU J.-M., "Safety management in a competitiveness context", *EUROSAFE Forum Proceedings*, 3–4 November, Paris, France, 2008.

[ROU 08b] ROUSSEAU J.-M., LARGIER A., *Conduire un diagnostic organisationnel par la recherche de facteurs pathogènes*, Techniques de l'Ingénieur, AG 1576, 2008.

[ROU 11] ROUSSEAU J.-M., ALENGRY P., JOURNÉ B. *et al.*, Les facteurs organisationnels et humains de la gestion des risques : idées reçues, idées déçues, IRSN Report, DSR no. 438, available at: www.irsn.fr, 2011.

[ROU 14] ROUSSEAU J-M., Faire du REX aujourd'hui : pourquoi ? comment ? – Repères pour un retour d'expérience événementiel source d'apprentissages, Report IRSN PSN-SRDS/2014-00019, 2014.

[ROU 16] ROUSSEAU J.-M., DECHY N., LARGIER A., "Maîtriser le changement pour maîtriser les risques : quand l'amélioration continue devient illusoire", *Congrès IMdR λμ20*, 11–13 October, Saint-Malo, France, 2016.

[ROU 18] ROUSSEAU J.-M., MONTMEAT A., BEN AYADI M. *et al.*, "Rex et digital : pas de bras, pas de big data !", *Congrès IMdR λμ21*, 11–13 October, Reims, France, 2018.

[SAG 93] SAGAN S., *The Limits of Safety*, Princeton University Press, Princeton, New Jersey, USA, 1993.

[SAN 17] SANTANA D., Manager la culture de sûreté : construction, représentations et usages de la "culture de sûreté" dans l'industrie nucléaire, PhD Thesis, Sciences Po, Paris, France, 2017.

[SEN 95] SENGE P., *La cinquième discipline : levier des organisations apprenantes*, Eyrolles, Paris, France, 1995.

[SGD 14] SGDSN, Plan national de réponse – Accident nucléaire ou radiologique majeur, Numéro 200/SGDSN/PSE/PSN Edition, 2014.

[SHO 84] SHORT J., "The social fabric at risk: toward the social transformation of risk analysis", *American Sociological Review*, vol. 49, no. 6, pp. 711–725, 1984.

[SIL 82] SILLS D.L., WOLF C.P., SHELANSKI V.B. (eds), *A Westview Special Study: Accident at Three Mile Island. The Human Dimensions*, Boulder, Colorado, USA, 1982.

[STA 05] STARBUCK W., FARJOUN M. (eds), *Organization at the Limit. Lessons from the Columbia Disaster*, Blackwell Publishing Ltd, Oxford, UK, 2005.

[STO 10] STOESSEL C., Décisions risquées et organisations à risques : autonomie au travail et reconnaissance sociale dans la conduite d'une industrie de process, PhD Thesis, CNAM, Paris, France, 2010.

[TAS 00] TASSET D., LABARTHE J.-P., "The impact on safety of computerized control room in nuclear power plants: the French experience on human factors with N4 series", *Proceedings of the Human Factors and Ergonomics Society Annual Meeting*, vol. 44, no. 22, pp. 815–818, 2000.

[THE 17] THELLIER S., Approche ergonomique de l'analyse des risques en radiothérapie : de l'analyse des modes de défaillances à la mise en discussion des modes de réussite, PhD Thesis, Conservatoire National des Arts et Métiers, Paris, France, 2017.

[TIL 16] TILLEMENT S., GENTIL S., "Entre arrangements et empêchements dans le nucléaire, une analyse par l'activité de la coordination au travail", *Sociologie et sociétés*, vol. 48, no. 1, pp. 117–142, 2016.

[TIL 19] TILLEMENT S., HAYES J., "Maintenance schedules as boundary objects for improved organizational reliability", *Cognition, Technology and Work*, vol. 21, pp. 497–515, 2019.

[TOS 12] TOSELLO M., LEVEQUE F., DUTILLIEU S. *et al.*, "Conditions for the successful integration of Human and Organizational Factors (HOF) in the nuclear safety analysis", *Work*, vol. 41, pp. 2656–2660, 2012.

[TUR 78] TURNER B., *Man-Made Disasters*, Wykeham Publications, London, UK, 1978.

[VAU 96] VAUGHAN D., *The Challenger Launch Decision. Risky Technology, Culture, and Deviance at NASA*, The Chicago University Press, Chicago, USA, 1996.

[VAU 97] VAUGHAN D., "The trickle-down effect: policy decisions, risky work, and the challenger tragedy", *California Management Review*, vol. 39, no. 2, pp. 80–102, 1997.

[VAU 99] VAUGHAN D., "The dark side of organizations: mistake, misconduct, and disaster", *Annual Review of Sociology*, vol. 25, pp. 271–305, 1999.

[VAU 11] VAUTIER, J.-F., TOSELLO, M., BARNABÉ, I. *et al.*, "Développement du réseau Facteurs Humains et Organisationnels (FOH) du CEA : un témoignage réflexif", in: Garrigou A., Jeffroy F. (eds), *L'ergonomie à la croisée des risques*, Actes du 46ème Congrès de la SELF, Paris France, 2011.

[VAU 18a] VAUTIER, J.-F., *La notion de configuration dans des approches systémiques pour appréhender la complexité*, Techniques de l'Ingénieur, Paris, France, 2018.

[VAU 18b] VAUTIER J.-F., DECHY N., COYE DE BRUNELIS T. *et al.*, "Benefits of systems thinking for a human and organizational factors approach to safety management", *Environment Systems and Decisions*, vol. 38, pp. 353–366, 2018.

[VAU 18c] VAUTIER J.-F., DECHY N., COYE DE BRUNELIS T. *et al.*, "Systemic characteristics of a human and organizational factors (HOF) approach of safety management", in: Barile S., Espejo R., Perko I. *et al.* (eds), *Cybernetics and Systems: Social and Business Decisions*, Routledge, London, UK, 2018.

[VAU 19a] VAUTIER J.-F., "Intégration des facteurs organisationnels et humains (FOH) dans la gestion des risques industriels : apport des couples dialogiques", *54th SELF Congress*, Tours, France, 2019.

[VAU 19b] VAUTIER J.-F., "Notion d'interaction, définitions et mises en perspective", in: Vautier J.-F. (ed.), *La notion d'interaction dans des approches systémiques pour appréhender la complexité*, Techniques de l'Ingénieur, Paris, France, 2019.

[VAU 21] VAUTIER J.-F., "Apport de la notion de situation de travail pour l'identification des facteurs explicatifs d'un évènement non-souhaité", *Proceedings of the 55th SELF Congress*, 11–13 January, Paris, France, 2021.

[VID 07] VIDAL-GOMEL C., ROGALSKI J., "La conceptualisation et la place des concepts pragmatiques dans l'activité professionnelle et le développement des compétences", *@ctivités*, vol. 4, no. 1, pp. 49–84, 2007.

[WEI 95] WEICK K., *Sensemaking in Organizations*, Sage Publications, Thousand Oaks, USA, 1995.

[WEI 01] WEICK K., SUTCLIFFE K., *Managing the Unexpected – Assuring High Performance in an Age of Complexity*, Jossey-Bass Publishers, Hoboken, New Jersey, USA, 2001.

[WIL 88] WILDAVSKY A., "Anticipation versus resilience", in: Wildavsky, A. (ed.), *Searching for Safety. Studies in Social Philosophy & Policy*, Transaction Publishers, Piscataway, New Jersey, USA, 1988.

[WIL 98] WILPERT, B., FAHLBRUCH, B., "Safety related interventions in inter-organizational fields", in: Hale A., Baram M. (eds), *Safety Management – The Challenge of Change*, Pergamon, Oxford, UK, 1998.

7

From Too Little to Too Much: The Impact of Big Data

The data needed for reliability calculations traditionally come from failure and maintenance records. Today, more and more, and sometimes from the design stage, sensors can be installed on equipment. They collect information from a large number of physical or operating parameters, making it possible to determine accurately the environmental conditions and the operating conditions of these devices, and monitor them in the context of conditional maintenance; the main goal is to send a diagnostics of the device's behavior, and then to predict its evolution. This chapter aims to show the processing opportunities offered by big data methods for estimating reliability, forecast a component's or system's performance and optimize maintenance. It also mentions the conditions required to benefit from this and the difficulties that may be encountered, as well as challenges for the near future.

7.1. Introduction

The context of risk analysis and dependability modeling will be greatly disrupted. The main reason for this is the arrival and eruption of new digital and technological tools (virtual assistants for collection, connected objects (Internet of Things (IoT)), Health and Usage Monitoring (HUMS) systems, industry and maintenance 4.0, and big data Natural Language Processing (NLP), etc.). By big data, we understand all the models for management, visualization, processing, analysis and interpretation of "massive" data (large volumes), or highly heterogeneous data (textual reports from intervention, digital data, photos of

---

Chapter written by André LANNOY and Emmanuel REMY.

equipment, etc.), or taken for the most part from HUMS system monitoring sensors or all three at once.

Operation feedback (collection and processing [LAN 94, LAN 11], see Chapter 2) is the first to be impacted and it will profoundly influence all other themes in risk management and dependability. These technological possibilities, with easier collection and new processing tools, will take us from a period of *too little data*, where modeling tends to be static, to a period of *too much data*, where a dynamic approach to reliability behavior can be envisaged. In reality, we are moving from this classical operation feedback with a low quantity of data, generally collected "by hand", to operation feedback where the data are massive, collected automatically, in line with multiple heterogeneous databases, of industrial data, context data, monitoring data and more physical and monitoring data, linked to conditional maintenance.

These new tools should provide us with a better knowledge of the usage profile (or operating profile), of degradations of any type, of failures, incidents and dysfunctions. This disruption can however call into question, in some ways, the approaches and models used at present, the interpretations of experts in analysis of the behavior of devices, and impact industrial practices taken from experience and accepted by safety authorities.

This chapter aims to demonstrate the potential consequences of this disruption on reliability studies as they are now. Many questions in fact can be asked.

– Won't everything and anything be collected? Won't we encounter problems of recording, data accessibility, data confidentiality, sharing and security?

– What methods and tools should be used to select a sample?

– How do we pre-process and validate this massive data in the classic sense of accuracy and relevance (see Chapter 2). Usual validation, which tends to be "manual", no longer seems imaginable because of the masses of data recorded. How to integrate human expertise? What will be the impact on the uncertainty of results?

– Will big data provide ease in estimating component reliability, for dynamic modeling of a system's reliability? Will it facilitate demonstration of interactions? Will it use "black box" models (e.g. neural networks, etc.), often closer to a systemic vision? How much trust should be given to the results?

– If big data will make it possible to understand a component or system's normal (or "nominal") behavior, will it be capable of processing operation feedback from rare events, of detecting weak signals?

– Can this facilitate prognostics, prediction, anticipation or, on the contrary, might not this make analysis and diagnostics more complex, because of the ease of collecting details?

This chapter refers especially to the [IMD 20] project carried out very recently with the support of 11 subscribers from different industrial sectors and to the communication [BOR 20], but also to a series of IMdR projects conducted since the 2010s with industry, particularly from the nuclear sector, focusing on the contributions of big data for processing weak signals [IMD 13b, JOU 14], and HUMS [IMD 17], as well as the contribution of natural language processing [IMD 13a, BLA 14]. It should also be indicated that the nuclear sector presents a technical singularity, considering the low number of failures and dysfunctions, on the one hand, but a large amount of operations and monitoring data collected, on the other hand. Studies carried out are generally half-way studies between small data (from classic operation feedback) and big data (big data from monitoring).

7.2. Toward a better understanding?

7.2.1. *New ways of collecting operation feedback*

Big data is not a revolution in reliability techniques. What is revolutionary is the acquisition of a new, more detailed, bigger, more heterogeneous operation feedback, which is becoming available for reliability analysis. Big data is characterized by the 3 V as follows:

– the volume, corresponding to a very large number of data saved in records, tables or files;

– the velocity, or speed of data creation, collection and sharing;

– the variety of data whether structured or unstructured, online or images or videos, human generated texts or machine generated readings.

Previously (and still today), reliability analysis depends on classic operation feedback data (Chapter 2). A new means of collecting data has emerged with the arrival of HUMS systems in the context of monitoring [EN 10]. Sensors have been installed in critical locations of systems with the aim of monitoring them as well as their components, to monitor health state then to estimate their reliability, indeed to predict their evolution, and above all to underline the (risk) factors that can jeopardize good functioning, to avoid failure and to optimize maintenance. The location of these sensors is generally defined as the design stage or even the re-design stage. A very large amount of data, particularly physical for data taken from sensors, is therefore collected, stored, and is becoming available. It is processed, analyzed and interpreted, using big data methods and tools.

The sections that follow will briefly present some methodological landmarks that can be advised for processing and interpreting massive data:

– pre-processing, which is essential, since it will guarantee trust in the results obtained;

– a better vision of the usage profile, which is also essential for reliability estimations, for understanding dysfunctions, or for probabilistic safety assessments (Chapter 3);

– machine learning algorithms for exploring data from a sample before processing or for classification;

– reliability estimation, which could call on covariate models (explained in Chapter 5), such as the Cox model, the logistic regression model, or neural networks, etc.;

– visualization of results to be interpreted by expert.

Big data is not something simple such as a technology layer that could just be added to draw benefit from it. To use big data, it makes sense to start a complete overhaul of the whole system (information system, data collection then data validation processes, processing with the help of dependability methods). Big data must be envisaged from design, to more easily put in place monitoring sensors that will make it possible to follow the behavior of components, their functioning and the operating parameters. The data collected will then essentially be lists, point data or time series. Data from situation analysis (the classic operation feedback data mentioned in Chapter 2) remain relevant, they are vital. Professional analysis of a problem situation (failure or incident) will always be needed. On the contrary, having more precise, more plentiful operating and monitoring data, thanks to big data, will be a plus compared to the current situation.

Nuclear power plants are already well-equipped with instruments and a large number of operating measurements taken from sensors have been available since the 1990s. The computer format of this data has long restricted its use and it has now been updated. Besides, nor was there the necessary calculation power, nor adequate mathematical algorithms.

Global investment is important: investment in monitoring sensors to obtain additional data, investment in information systems adapted to big data, investment in new data management skills, data and analysis and processing skills, a need for dialogue between data experts (data scientists), reliability experts and operators to obtain and interpret results. It is therefore necessary that this investment merits its cost.

7.2.2. *The importance of pre-processing and validation*

This point has already been mentioned in Chapter 2, where it pertained to judging the accuracy (at the moment of collection) and the relevance (at the moment of analysis) of classic operation feedback data (see also [PET 99]). However, the context is changing. Tomorrow, we will find ourselves faced with millions of data from monitoring. Pre-processing by hand will no longer be thinkable.

Like any classical process, the machine learning process can also be written: data + models → forecasting. The quality of results will depend on data, its volume and models used. It can immediately be understood that the quality of the results depends first on the quality of the data, that trust in forecasts will depend mainly on this quality and that evaluation of this quality is a challenge for the reliability specialist faced with the validation analysis of millions of raw data before any use or processing. This quality is more even present during collection than before. Indeed, the automated character and volumes of the data collected invite us to defer checks at the moment of analysis. In this sense, there is a need to develop further monitoring procedures, analysis of the situation and validation during collection. This is very often neglected by decision makers and even by reliability specialists experts, even though it is the condition *sine qua non* for obtaining representative, relevant and robust results.

"Data" are not neutral. They are the result of a process of analyzing and technically and socially constructing reality (e.g. to what dimensions of reality do we pay attention? On which part of a device should a sensor be placed? To monitor what and to collect what type of data, in what format? What are the biases introduced by measurement and detection instruments?).

The first condition is that the reliability expert has a good knowledge of the data and their collection and analysis process. Three domains [LAW 17] should be considered:

i) The accessibility of the data: do they exist? Really? In what format (manual, textual, outdated formats)? Are they private or public? Are there ethical constraints? Limits on access? After these points have been examined, the data can be manipulated.

ii) The quality of the raw data and how representative it is: are there missing values? Outlying values? Have errors from sensors and their deviations been considered? How reliable are the sensors? Is "manual" data validated? Are the units used correctly? At this level, descriptive analyses or graph visualizations can be made; deciders and reliability experts can already appreciate the limits or strengths of their data, and know, if only intuitively, what more data processing might contribute.

iii) The relevance of the data: this is the main preoccupation of the reliability specialist for so-called "classic" operation feedback (Chapter 2): are the data adapted to the problem posed and to the context? Can they be used for forecasting? What tasks should be carried out to clean them, to validate them, to ensure their relevance for the problem posed? At this level, reliability specialists, from their skills and experience, can focus on the quality of the data available, then proceed to data analyses.

It is well understood that this process is weighty and difficult, and therefore costly, that it is not automatic and that it requires great expertise for automating the collection and processing of data. That is the very basis of the trust that we will be able to attribute to the forecasting results. Acquisition of poor results may be the consequence of a poor pre-processing of raw data, just as much as of an inappropriate or incorrect use of statistical algorithms. On the contrary, a good pre-processing will surely guarantee trust and will help businesses to develop representative operation feedback databases correctly.

7.2.3. *A more accurate vision of the usage profile*

The usage profile, also called the operating profile or the working profile, is important for reliability studies on systems (especially for probabilistic safety evaluations, see Chapter 3) and for studies on structures (Chapter 4). Big data is an opportunity to better know the usage profile. It makes it possible, thanks to information accessible on heterogeneous databases, to better identify and evaluate different periods of operation (full load, cold shutdown, load variation, etc.), in frequency and in duration, for a system, structure or component in operation. It will thus be possible to estimate an overall operating time or the time passed in each of the operating periods cited above. It is in fact important to know the operating conditions, which can be divided into categories, for example:

– Category 1: normal operating transients, starting – shutdown of a plant, thermal transients, load variations, etc., corresponding to normal functioning.

– Category 2: the loss of external power supplies, etc., leading to a plant shutdown followed by immediate restarting.

– Category 3: steam generator tube rupture, etc., followed by a plant shutdown of variable length, relatively long.

– Category 4: breach or rupture on the primary circuit, etc., leading to a shutdown of indeterminate length.

Note that in the design stage, the usage profile is defined at the outset, either as the n-1 system profile, often from analysis of past operation feedback, or by the expression of specifications planning a new usage of the system n.

7.2.4. *Toward big data methods*

The first task that can be engaged in on the big data sample is similar to a screening approach, that is, to selecting, sorting, and filtering information. This stage is achieved using data mining or text mining (when the data are textual; it should be noted that the processing of increasingly textual operation feedback is becoming vital, even in reliability), using statistical analysis methods. The sample is then explored at the outset using visualization, classification (supervised or unsupervised) into families, a search for correlations or atypical points, hypothesis statistical tests, etc., to give orientations on the modeling to be adopted. This exploration presents the following benefits:

– A better knowledge of functioning and dysfunctioning, and therefore a better understanding of failures, by monitoring a very broad spectrum of parameters.

– A classification of the operating situations observed; the arrival of a new, unclassifiable situation is perhaps the sign of a weak signal (also called early warning); we can note that the current operating life will be recorded, including weak signals (which will not therefore be recognized as such, since additional analysis is needed); these weak signals can lead to extreme events with very low probability, and very damaging consequences; there are therefore very few extreme events recorded in the databases; the detection of evolutions in parameters, signals taken from sensors and trends.

– A better visibility on the parameters to be monitored as a priority, and especially those that can lead to a degradation or failure; *HUMS* sensors placed in a monitoring system essentially measure physical parameters; it is then possible to have more data over time to determine a degradation kinetics or a residual (or remaining) life duration.

– From raw data extracted for each problem size, statistical characteristics (called features) are defined, which may be an average, a median, a quantile, a dispersion, a proportion of time passed in a particular configuration, etc., and which synthesize the information contained in the data; these features are the input data for the machine learning model.

– Using machine learning to highlight the correlation statistics between different parameters (internal or external, context, environment) and to deduce interactions from them; the reliability or physical state are linked to some of these interactions and to influencing parameters, which are called covariates, contributing concurrently to the degradation then to the failure of the system's functions.

– When the machine learning algorithm is built and validated (generally by its results and by expertise), then forecasts can be made, since the influencing parameters are determined. This forecasting model is a statistical model containing some of the parameters considered important initially, but also perhaps forgotten parameters the influence of which was not suspected and which have been identified using machine learning algorithms. The forecasting will be even better if the phenomena can be explained and can thus contribute to improving a potential, existing physical model.

7.2.5. *Reliability approaches*

After using algorithms of big data, reliability models can be used in a second stage. Some machine learning algorithms, such as the one presented in [LEE 18] and introduced in Chapter 5, can provide reliability predictions without relying on a classic reliability modeling approach. Figure 7.1 taken from [GOR 12] and with additions by [IMD 20] presents reliability approaches to date. It classifies three domains of approaches:

– The classic reliability approach, which relies on a probabilistic data modeling (including some models and tools presented in Chapter 5), including graphic methods, data analysis methods, regressions (in particular logistic regression), classic methods for estimating the parameters of a reliability distribution (such as the Weibull distribution for ageing) and the impact of maintenance on reliability, models with covariates, time series models (such as auto regressive moving average [ARMA]); degradation analyses with the help of processes such as gamma or Lévy or Wiener can also be cited, they can forecast a residual life. Probabilistic networks (in particular Bayesian networks) could be useful for modeling the evolution of a degradation in a complex context and in respect of physical laws and consequently for proposing an optimized maintenance program. They have moreover the advantage of providing a graphical view. We note the recent thesis [SHA 20] on the "health" management of complex systems. The thesis focuses on dynamic probabilistic graphical models making it possible to diagnose a system's state of health, then the prognosis, degradation evolution and the estimation of the residual life. It can be thought that covariate models will certainly be favored in predictive reliability studies, in operation or in extended operation. One representative of these models is the Cox model called proportional hazard model (PHM) and presented in section 5.4.3 of Chapter 5 [COX 72]. It makes it possible to consider the impact of the operating time and of covariates (eventually time-dependent), assumed to be independent of one another in calculating the failure rate. The regression coefficients βi of the model measure the impact of each covariate i on this rate, and they are generally determined by the maximum likelihood method (see section 5.4.3). The explanatory covariates act multiplicatively on the failure rate, hence the name "proportional hazards"; if a βi factor is distinguished by its impact, it is helpful to return to the physical model, if it exists, to understand the physical phenomena and to amend this physical model.

– Approaches using physical models, corresponding to numerical simulation codes, to analysis and degradation kinetics models, and to physical-reliability models from Chapter 4; these are essentially tools used for structural reliability for which we wish to estimate the evolution of a damaging process and the RUL; this practice is used particularly in thermal fatigue (finite element codes, TLAA [time limited ageing analysis], accounting of situations, etc.).

– Finally, big data processing driven approaches corresponding to artificial intelligence algorithms making it possible to demonstrate correlations between variables, to detect "forgotten" variables, to demonstrate their importance, to build algorithms making it possible to reproduce observed situations described by the data; expert systems based on rules were used in the years 1980–1990, but they proved difficult to operate in the presence of plentiful heterogeneous data and were not convincing; we can cite neural networks, which are black boxes, thus may be difficult to validate and be accepted with difficulty by engineers and safety authorities, no matter how well they perform; it can be thought that a "hybrid" approach where any available information (data, data-driven methods, reliability models, physical models and expertise) would be aggregated would lead to a consistent approach, containing substantial lessons, efficient, guaranteeing the trust of decision makers and end-users.

Machine learning techniques make it possible to use data and they rely on algorithms without probabilistic modeling. The output of the algorithm learns from the experience contained in the data. Performances increase with the data mass. We distinguish supervised machine learning, which includes classification and regression algorithms. Unsupervised machine learning is focused on the search for similarities (clusters, patterns), links or associations between objects, or there are attempts to reduce the dimensions of a subject to deduce its main properties.

Machine learning sometimes highlights the variables whose importance had not been noticed before. This can only facilitate physical understanding and we also think that recourse to physical models is essential. The physical model can in fact help reliability experts in their analysis and interpretation of correlations detected during the data mining stage (section 7.2.3) of their decision.

Big data must primarily be an aid for demonstrating influencing parameters and risk factors, then an aid for physical understanding of phenomena. It gives us the opportunity to improve our approaches and modeling, not only of systems in operation but also future systems currently being designed. This is what is meant by the hybrid approach in the technical documentation. Analysis by the reliability expert makes it possible to integrate the correlations and findings observed from data analysis in the physical model.

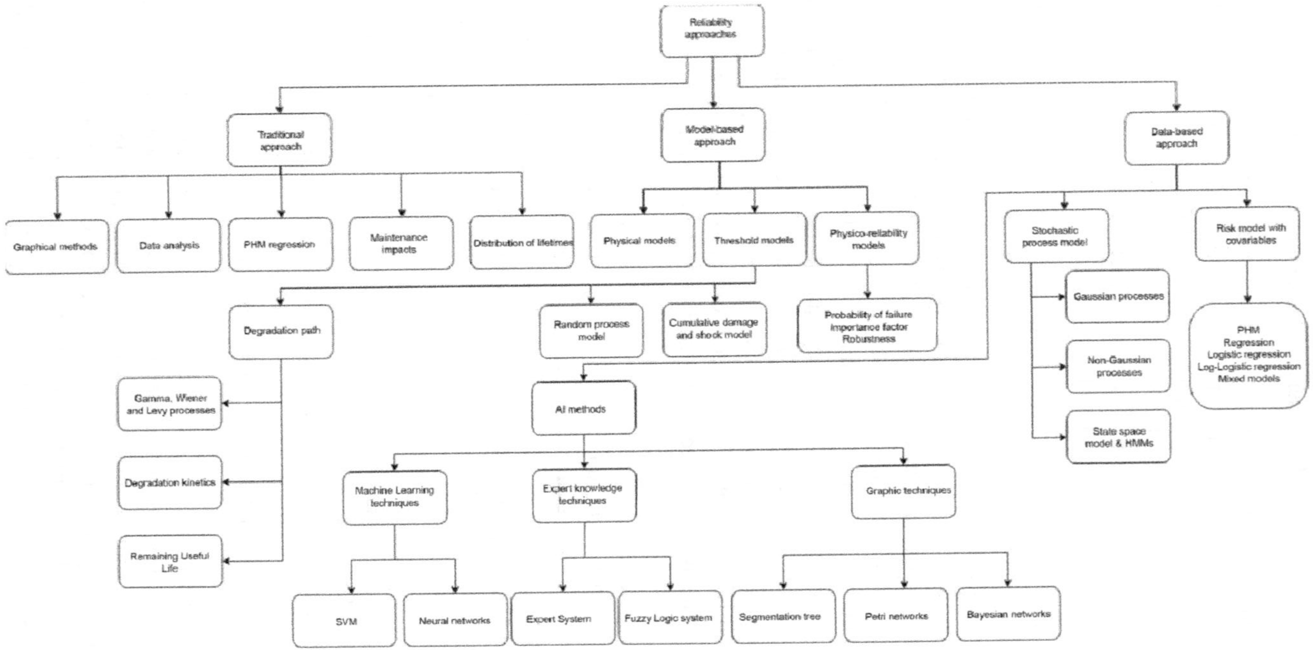

Figure 7.1. *Reliability Approaches (legend: AI = Artificial Intelligence methods; HMMs = Hidden Markov Models; PHM = Proportional Hazard Models; SVM = Support Vector Machine)*

Table 7.1 briefly compares the classic "small data" reliability approach to the big data approach on some features.

Classic reliability approach	Big data reliability approach
Medium investment	Substantial investment
Mathematical reliability modeling approach	Machine learning and reliability approach
Failure (loss of function) and maintenance data	Heterogeneous, internal and external data
Small samples	Big data make pre-processing and analysis difficult
Uncertainties, simulations	Short-term predictions

Table 7.1. *Comparison of classical and big data approaches in reliability*

What characterizes big data and differentiates it from small data?

– Observations of failure, or of degradation, in lesser numbers, or even very low numbers, because potential failure has been detected and maintenance has been involved positively to delay or eliminate it; this is the variable to be explained (the rows of the database matrix) for which there is, in the end, very little information, perhaps even less than in small data.

– An explosion in the number of covariates or features, because of HUMS, among others, which makes it possible to collect a great deal of information or parameters potentially linked to the state of health of the entity being monitored, since all this information is not necessarily independent, since it is generally at the design or redesign stage that the decision is made to equip installations with monitoring sensors; these are the explanatory variables, so the columns of the database matrix.

In the context of reliability, there is not therefore "big data" on both dimensions of the database matrix, and this is a difficulty that may be encountered in the industrial world, unlike in the Internet, commercial or medical worlds for example, which have many more row observations. It is a half-way point between small data on the rows and big data on the columns. Another problem may appear: the algorithm may not have enough data for the learning stage, which is vital for machine learning techniques.

7.2.6. *A posteriori processing or visualization*

This stage consists of analyzing results and their use. Knowledge should be accessible to experts then to decision-makers. Visualization of results facilitates, through its graphical aspect, the interaction with expertise, discussion, improvement, then decision-making and better perception. It facilitates interpretation and judgement on the relevance of results, it reflects trust and finally it involves different skills in the organization, from collection to management, thus sharing knowledge.

We recall that visualization using descriptive analysis is useful too in the pre-processing phase (section 7.2.1), for example for catching extreme or outlying data.

A few caveats can be stated. When the results are produced, it makes sense to explain these results physically, to properly understand and potentially propose a new physical modeling.

The "black box" type model is not generally accepted by the engineer, the decision-makers and the safety authorities: the results of such a model can however be interesting and they must not be neglected, it is however essential to have them analyzed by field experts.

Big data, such as moreover classical or small data approaches, is limited when there is a question of forecasting the behavior of a new system being designed. Either the data do not exist, or they are too old, or they do not correspond to the operation, maintenance or environment conditions. Big data is probably lacking for extreme events with low probability, unless there is a very significant history. Similarly it may also be limited or faulty when new sensors are introduced into an existing system. New covariates are thus introduced, new results will be obtained. Consequently the model will have to be modified.

7.3. Diagnostics and prognostics

7.3.1. *Diagnostics*

Standard EN 13306 gives the following definition of diagnosis (called also health status): actions taken for fault recognition, fault localization and cause identification. Diagnostics therefore creates a reliability assessment of the system or the component. This assessment is a snapshot of the state of the system or component at a moment of observation. It specifies the life duration already consumed or a usage factor. This is an unavoidable task if we wish to improve reliability or make prognostics.

In the context of monitoring their plant, the operator of an industrial asset checks periodically if it is in good working order and establishes this assessment periodically. They counts the failures observed, identifies the direct or apparent causes, and analyzes in-depth causes and failure modes. They check in particular the nature of potential defects and degradations, their location, their extent and monitors their importance and propagation over time. These measures are used to characterize the evolution of degradations. They correspond to a measurable effect (see Figure 2.2 from Chapter 2): the consequence of a degradation mechanism that can be measured; this may, for example, be a length or a depth of crack, a loss of thickness, a vibration level, a flow, a wear out, etc. This quantitative data, called inspection–control data, are often recorded in specific databases.

Degradation is generally triggered by a physical phenomenon often called a degradation mechanism. Degradation (irreversible evolution of one or more features of a good, linked to time, to the duration of use or to an external cause, according to standard EN 13306) is not a loss of function, but an alteration of the function. With big data analysis preventing failures, the failures should decrease in number and analysis of degradation kinetics should become a priority concern. It might be thought that linking HUMS databases, recorded continuously in operation feedback databases (whereas previously, they were analyzed periodically), with the inspection-control database, should facilitate physical understanding of the degradation process through the contribution of explanatory variables and through searching for correlations between these variables and degradation measurements.

The arrival of big data and machine learning algorithms should make it possible to better understand breakdowns, by demonstrating the influencing factors/parameters, as seen previously (section 7.2.3), and finally to better avoid failures, the number of which should be reduced. In fact, monitoring the evolution of degradation becomes an important task for the reliability specialist and the maintenance engineer [NIK 02, LOR 10].

For example, Chabridon *et al.* [CHA 20] present a methodology for identifying influencing factors then for forecasting the degradation kinetics of an industrial component using a supervised machine learning tool.

The aim is to construct a model whose role is to demonstrate the evolution of the degradation with input observations (usage profile, monitoring of operation and context parameters: geometry of equipment, material properties, presence of defects, etc.). The measurable effect can be conveyed by a function of the type:

measurable effect = f (operating time, explanatory factors/parameters).

The input data are the collected, quantitative data from operation feedback and monitoring, but also other data from design, from maintenance, from equipment, etc. Operating times and other explanatory variables are recorded during operation. The measurable effect, which is the output data that we seek to explain, is measured at precise moments during inspection-control campaigns. Graphical representation of this measurable effect makes it possible to monitor the evolution of the degradation and to predict, using "extrapolation", the date of a preventive maintenance, that is, the delay before reaching of a threshold or limit that triggers a preventive maintenance intervention. This limit level is fixed by technological criteria or safety criteria. We also seek to determine the remaining useful life (RUL).

In [CHA 20], it is a multiple linear regression model that is used. After reducing the number of features to keep only the most relevant ones, construction of the model passes through two stages: a training step and then a validation step. Several different models have been developed and tested. The search for the model thought to be most relevant was finally made by the reliability expert, on the basis of the presence of known influencing parameters on which the operator can act to limit the evolution kinetic of the degradation mechanism.

7.3.2. *The prognostics*

The prognostics relies, on the one hand, on the diagnostics and so on the data history, and, on the other hand, on future operating conditions (new usage profile, more aggressive preventive maintenance, design modification, etc.). Standard EN 13306 does not define prognostics, but defines improvement as the set of technical, administrative and management measures intended to improve the dependability of a good without changing its required function.

The prognostics can be based on physical models, data-driven models or hybrid models combining both approaches.

Data-driven models are not based on operating models. As we saw previously, approaches are generally made using classification or by regression. Many algorithms are available. They generally require a training dataset, considered as a reference, and a validation (or test) dataset basis. These datasets contain, among others, raw monitoring data features. Once the forecasting model has been built and chosen, the user will use it by introducing as input the values of the features considered plausible in the future or reflecting the future operating conditions. Projection (extrapolation or forecasting) can only be achieved at a short distance from the perimeter of study of the available data that were used to train the model.

Prognostics is integrated into predictive maintenance and it is a concern for the maintenance engineer. According to standard EN 13306, predictive maintenance is a conditional maintenance carried out following the predictions extrapolated from analysis and evaluation of significant parameters of the degradation of a good. It therefore considers monitoring, then an "extrapolated" prediction, also called "proactive" or, sometimes, exceptional maintenance.

It essentially means predicting the moment at which the system or device will cease to fulfil its function. It is necessary to predict future performances, the evolution of degradations, the moment from which the system will deviate from its normal operating conditions. At the end of the 1990s, this task was accomplished in the context of lifecycle management (LCM) [SLI 03]. Today, we speak of prognosis and health management (PHM) or structural health management (SHM). This means tracking systematically whatever may deteriorate behavior, identifying components that will cease to function. Reliability becomes the health criteria of the installation or its equipment. The information sought is the distribution of the time remaining before failure (the distribution of the RUL), estimated from past operation feedback and monitoring data. This distribution also makes it possible to evaluate the probability of functioning without failure or the time of the next preventive maintenance to carry out.

Conditional maintenance bases its strategy on monitoring measurements from devices. It means taking action only when it is required. This makes it possible to avoid unnecessary preventive (intrusive) maintenance operations. Prognostics aims to predict the most probable failures.

7.3.3. *Classical reliability models for prognostics*

These models remain relevant. They are based on probabilistic models and the estimation of these models' parameters, for example the parameters of a Weibull distribution estimated using the maximum likelihood method (Chapter 5). The evolution of reliability depends on the operating conditions, exogenous conditions and on other factors or explanatory parameters. This estimation will therefore be improved by:

– considering these explanatory factors;

– a better knowledge of the usage profile;

– an anticipation of failures by studying degradation kinetics.

We distinguish three main stages of using the classic reliability approach for a predictive maintenance problem:

i) Data acquisition: classic event data, historical data and HUMS data taken from monitoring, design data (geometries, material, etc.), meteorological data, etc..

ii) An analysis, modeling and data processing stage for estimating reliability, where we find the evolution of the degradation, the regression, physical models, classical models such as the Weibull distribution, etc.; these models make it possible to make a diagnostics followed by a prediction.

iii) This prediction is corrected by expertise (prior information) or by other information, during the third stage; we can give a value (a weight) to these different sources of information, the aim being to reduce the uncertainty level, for example via a Bayesian approach; actions or maintenance options are then defined, which are optimized by a cost model (or asset management model), which makes it possible to establish a predictive maintenance policy.

It can be noted that many data enrich the approach and the uncertainty will be reduced. It will be possible to monitor a system or component individually, more easily, throughout its lifecycle. The objectives will be optimized.

7.4. Trust

The use of big data should provide significant assistance, making it possible to demonstrate the risk factor parameters, to weigh their influence and to anticipate the future. These algorithms are useful for detecting these influencing parameters for anticipating and forecasting.

This is already done very well with "simple" standard Cox type models or accelerated life models, and so without a big data type algorithm. In this classical usage, it is the reliability expert who is first in line to direct the choice of candidate covariates, by favoring those easily accessible and thought to be influential. With big data, it is the algorithm that will designate the covariates with, probably, more effectiveness on condition that the right input variables have been well chosen. However, it is always possible to question these results demonstrated by the machine or rather by the algorithms, and not by the human expert. There will only be trust in the forecasting results if the results are explainable and explained.

We can already affirm that a low investment by the business in its information system, a careless pre-processing, a resolution made by black box models, the absence of involvement by a decision-maker, reliability experts or field experts, are a number of arguments militating for distrust of the results obtained.

Unlike big data (which is high in volume), a quality pre-processing, the highlighting of important parameters and forecasts that are verified, explainable algorithms, the evaluation of uncertainties, a physical interpretation and a comparison with the physical model, recourse to reliability experts for methods and estimations, and to field experts for interpretation, are a number of strong arguments for trust in the results.

The great advantage of this big data approach is quite clearly that of working on numerous, big, continuously recorded data obtained in real operating conditions – environment – maintenance of systems, which reliability tests (rarely carried out in real operating conditions) cannot take into account, or probabilistic reliability models generally simplified and supplied by input data, the main fault of which may be its low quantity. These "theoretical" reliability models could be called into question in coming years over their use in the operation–maintenance stage. On the contrary, physical models should experience rejuvenation: considering models with new influencing parameters, considering real operating conditions, less uncertain data, which should improve physical understanding of the phenomena and reduce the occurrence of dysfunction. It might also be thought that the use of this approach will be favorable to the development of models with organizational and human factors, through demonstrating important parameters, interactions and the construction of scenarios (see Chapter 6).

7.5. Conclusion and perspectives

The big data approach is a new tool to assist the reliability specialist, a precious tool, which does not however replace experts and their ability to understand, interpret and make decisions. The target objectives and the results envisaged are eloquent: preventing failures, avoiding failures, highlighting influencing parameters, measuring their importance, facilitating sensitivity analysis, improving safety, optimizing maintenance, putting in place support logistics, modifying the organization of maintenance and forecasting their future state. We can have faith in big data algorithms, on condition that their results are explainable and explained.

The conditions for a big data application to succeed depend on managerial objectives and methods for implementing or renewing operation feedback, on an effective information system, on equipping the plant with monitoring sensors at the outset and on data quality being guaranteed from collection. They therefore depend on management and its level of engagement. This in fact involves significant and costly effort. Other difficulties seem to us to be situated more in pre-processing, which is essential and time consuming, but which should be carried out seriously, since it is on the quality of pre-processing that the results and decisions that will be taken should depend. Indeed, the contributions of automating data collection and data volumes invite us to transfer the problem of data validation to the processing stage.

The big data tool is an "increased intelligence" tool that will make it possible to process flows and masses of data, which will reinforce the dynamic character and temporal relevance of estimations. It is an aid for the reliability expert; it gives indications that the reliability expert might not perhaps have been able to demonstrate "manually". Reliability missions will be modified: the reliability expert will also have to be a data expert, data scientist, familiar with their field of application, open to the physics and operating modes of organizations and individuals located in diverse work configurations. Roles and responsibilities will be redistributed as well as the organization of work to carry out the activities of data collection, validation and interpretation of results.

It makes sense always to interpret and explain the results obtained and, from our point of view, there is a need for recourse to use of physical models and to analyses of human and organizational factors. The development of hybrid approaches is certainly an essential need and will be a theme for R&D in the coming years.

The sector of the nuclear production of electricity has been seized/gripped by this new technology for 15 years. However, in this domain, dysfunctions are few, data are very heterogeneous and many are textual, and databases were still multiple and separated not long ago. Big data can aid dialogue between databases and the defragmentation of data stores. The outlook seems clear. The approach, models and algorithms now seem mature, and some convincing applications exist in the nuclear industry. There is a need for us to follow up and develop applications, resolve them, evaluate their robustness, industrialize them and emphasize any potential difficulties linked to the data, at pre-processing, at interpretation of the results and when making decisions. It seems vital to us that the professions of risk, operation feedback, reliability, maintenance and safety seize this new opportunity while still being cautious on the conditions for success that must be brought together, the means to invest, the potential risks run and biases induced.

7.6. References

[BLA 14] BLATTER C., RAYNAL C., *Méthodes d'analyse textuelle pour l'interprétation des REX humains, organisationnels et techniques*, Congrès λμ 19, Dijon, 2014.

[BOR 20] BORDEAU V., DEBACHE G., ESCUDIÉ F., LE LOC M., *Big Data in Reliability*, Conference λμ 22, 2020–2021.

[CHA 20] CHABRIDON V., DERI E., REMY E. *et al.*, *Apprentissage statistique pour la prévision de la dégradation d'un composant industriel*, Online conference λμ 22, October 2020 to January 2021.

[COX 72] COX D.R., "Regression models and life tables (with discussion)", *Journal of the Royal Statistical Society, Series B*, vol. 34, pp. 187–220, 1972.

[EN 10] EN 13306: 2010, *Terminology of Maintenance,* 2nd ed., CEN, Brussels, 2010.

[ESR 20] ESREDA, "Advances in reliability, risk and safety analysis with big data", JRC Conference and Workshop Report, in : FERREIRA L., SIMOLA K., KOPUSTINSKOS V. (eds), *Proceedings of the 57th ESReDA Seminar, Technical University of Valencia*, 23–24 October 2019, 2020.

[FAU 19] FAURIAT W., Prognostics and health management: data-driven methods and decision under uncertainty. Integration of condition monitoring data and uncertainty reduction, Safety and Risk Team, CentraleSupélec, available at: http://gst-mi.fr/Coppermine/albums/userpics/10005/Fauriat -Prognosis-and-health-management.pdf, 2019.

[GOR 12] GORJANIAN J.N., "Asset health prediction using the explicit hazard model", Philosophy Thesis, Science and Engineering Faculty Queensland University of Technology, available at: https://eprints.qut.edu.au/57314/1/Nima_Gorjian_Jolfaei_Thesis.pdf, 2012.

[IMD 13a] IMDR P10-5, Méthodes d'analyse textuelle pour l'interprétation des REX humains, organisationnels et techniques, Maîtrise des Risques, ESTP University campus, Cachan, 2013.

[IMD 13b] IMDR P12-1, Détection et pertinence d'un signal faible dans le traitement d'un retour d'expérience, Maîtrise des Risques, ESTP University campus, Cachan, 2013.

[IMD 17] IMDR P15-2, Health and Usage Monitoring Systems (HUMS), Maîtrise des Risques, ESTP University campus, Cachan, 2017.

[IMD 20] IMDR P17-4 (2020), Big Data in Reliability, T3 – Impact du Big Data sur l'estimation de la fiabilité d'un composant, d'une fonction ou d'un système, Maîtrise des Risques, ESTP University campus, Cachan, 2020.

[JOU 14] JOUNIAUX P., HALIDA D., DECHY N. *et al.*, *Détection, pertinence et amplification des signaux faibles dans le traitement du retour d'expérience*, Congrès λμ 19, Dijon, 2014.

[LAN 94] LANNOY A., PROCACCIA H., *Méthodes avancées de traitement et d'analyse de banques de données de retour d'expérience*, Collection de la Direction des Études et Recherches d'Électricité de France, Eyrolles, Paris, 1994.

[LAN 11] LANNOY A., *Retour d'expérience technique*, Techniques de l'Ingénieur, SE1041, 2011.

[LAW 17] LAWRENCE N.D., *Data Readiness Levels*, Amazon Research Cambridge and University of Sheffield, available at: http://inverseprobability.com/publications/data-readiness-levels.html, 2017.

[LEE 18] LEE C., ZAME W.R., YOON J. *et al.*, "Deephit: a deep learning approach to survival analysis with competing risks", *Thirty-Second AAAI Conference on Artificial Intelligence*, New-Orleans, February 2–7, 2018.

[LET 16] LETOT C., DEHOMBREUX P., *Comparaison des processus gamma et Wiener pour modéliser la dégradation de voies ferroviaires*, Congrès λμ 20, Saint-Malo, 2016.

[LOR 10] LORTON A., FOULADIRAD M., GRALL A., *Pronostic d'un système complexe : impact de l'information disponible*, Congrès λμ 17, La Rochelle, 2010.

[NIK 02] NIKULIN M., BAGDONAVICIUS V., *Accelerated Life Models, Modelling and Statistical Analysis*, Chapman & Hall/ CRC Press, Boca Raton, 2002.

[PET 99] PETTERSSON L., MOSS B., LANNOY A. *et al*, Handbook on quality of reliability data, An ESReDA working group report, Det Norske Veritas, 1999.

[SHA 20] SHAHIN K.I., Modèle graphique probabiliste appliqué au diagnostic de l'état de santé des systèmes, au pronostic et à l'estimation de la durée de vie résiduelle, Thesis, University of Lorraine, Nancy, 2020.

[SLI 03] SLITER G., *Life Cycle Management in the US Nuclear Power Industry,* SMIRT 2017, Prague, 2003.

8

Conclusions and Prospects

> It is for machinery as it is for the systems in which men place themselves, they call for more innovation and at the same time for more security. Of course, this cannot always be achieved. Technological progress, like social progress for that matter, is accompanied by new risks that we do not always know how to combat effectively, because of their novelty. Reliability or the 'science of trust' is born of such preoccupations. How can these two imperatives best be satisfied: innovation and security? There are technical means: improving the equipment's own qualities; there are also design methods: choosing optimally, if possible, new structures for the elements of this machinery, … . Recently, there has been a mathematical theory of the reliability of technical systems, … .
>
> Arnold Kaufmann, 1969, Polytechnic Institute of Grenoble

Reliability is taking an increasingly important place in all industrial sectors and even more in the nuclear energy sector. It is a continuous improvement approach. A review of evolutions since 1970 summarizes the importance of the cumulated knowledge to build up trust in the operating assets. After having summarized the objective of reliability, we cite the main operational R&D prospects, which are emerging from reliability analyses and which can be implemented in the coming years. A restricted list of references is presented as starting points for new progress.

Chapter written by André LANNOY.

8.1. Nuclear power plants and the progress of reliability

How much progress has been made, how many scientific advances since the 1970s! At this period, reliability was integrated into operational research. It was essentially used in the military sector, especially the navy sector and in designing electronic components for air and defense systems. There was a focus on exponential distribution failure rates, on series-parallel reliability diagrams, on failure mode and effects analysis and in technical trust.

Everything changed in the 1970s with major emerging risk concerns, of which the Flixborough disaster (the explosion of an unconfined vapor cloud of cyclohexane) raised awareness, with the development of new technologies, and in particular the development of nuclear power plants, for which security/safety requirements were imposed. The consideration of potential accidents from the design stage reached an unprecedented level in industry. This era, after 1970, was also a period where concerns about consumer behavior were considered. It led to improving the quality and performances of products and equipment. Reliability techniques were then introduced into the mass production sectors.

The 1970s, well before the coupling of the PWR power plant at Fessenheim (in France) in 1978, saw the beginning of safety and reliability studies in nuclear power plants, giving rise to an engineering science, called *dependability*. Beside the initial security/safety concerns, it primarily includes not only reliability but also availability, maintainability and the impact of maintenance and support logistics. Since the first connections at Fessenheim in 1978, dependability came to rely gradually on the capitalization of the operational experience feedback.

A great deal of research was then undertaken in the nuclear energy sector by designers, operators, logistic providers and safety authorities. However, estimating reliability requires data. The reliability approach includes statistical studies of situations or failures occurred and requires recourses of analysis methods and models to make it possible to evaluate the reliability of nuclear power plants and their equipment, and thus to contribute to improving their safety and availability. It is not strange, therefore, that the first studies to involve operational experience feedback dated since 1974. Then reliability methods of components since the first operation feedback were obtained, which was followed by human reliability analysis in 1979, systems reliability methods for Probabilistic Safety Assessment by 1980s, organizational reliability and structural reliability methods in 1986, the optimization of maintenance at the beginning of the 1990s, the integration of operation feedback in the design of new installations in 1994, ageing studies by the end of the 1990s, durability and management of industrial assets in 2004. Finally, the predictive maintenance and extension of life duration started by the 2010s, monitoring and HUMS systems in 2010 and the contribution of big data from 2015.

In all countries, generations of researchers looked into these problems and made progress in reliability, making it much complete and complex, with more concepts, mathematical models, identification, evaluation, forecasting and decision support methods. The goals are always the same: improving the performance of systems and components related to safety and optimizing their availability and maintenance. Reliability in 2020 is no longer that of the 1970s, and this is the reason why we have written this state-of-the-art and have decided to trace some prospects as described above. Researchers in nuclear energy industry owe a great deal to managers, to designers and operators of nuclear installations, to authorities, experts and academic research infrastructures and assets.

The very first decision taken from 1974, that is, reflecting on operation feedback for building and having real data, was fundamental. Without data, there is no system reliability engineering. Without data of high quality, no results of high quality are possible, neither safety nor trustfulness. Operation feedback is strategic for nuclear business and precious.

The data quality has always been a difficult problem for the nuclear industry. Since the early beginning, power plants have been designed with many precautions, very important security margins and constant demand of surveillance and maintenance. Failure and event data are still few even today after more than 40 operation-years. Reliability experts and engineers are still forced to work with small size samples of failures and a substantial amount of incomplete data. One may think that the situation would change positively with the arrival of Health and Usage Monitoring Systems and big data. Data would become massive and essentially be issued from monitoring and sensing activities. That would allow us to better know the operating conditions of equipment, and perhaps to better understand the physical phenomena and the fundamental causes of failures. But it could be argued that if failures become better explained and predicted because of a comprehensive knowledge of the influencing parameters, there will then be even fewer failure and events data than previously. The paradox is that, thanks to sensors, we should develop ever more effective predictive models with lower and lower volumes of data. The main weakness of reliability analysis will lie in this lack of input data and their high statistical uncertainty level due to data rarefication. This means that sensitivity analysis of input data is essential, since it can give us a measure of trust, of robustness of results and of the credibility of the reliability engineering sciences methodologies, models and results.

The second major fact is an opportunity. The designers, operators, maintenance engineers and authorities have identified and posed real problems, encountered in the activities of nuclear power plants, in the design stage as in the operating stage. The needs were well identified, the hypotheses were clear, the input data well-specified, without any doubt: well-posed questions and well-formulated hypotheses. It was not necessary to invent or imagine problems.

The third major fact is also an opportunity to profit from the expertise and the skills of interpretation and comprehension of experts and engineers in the field. Thus, it makes it possible to validate approaches and methods to transfer them more easily for practical applications, in power plants engineering or in field engineering. We can emphasize on the fact that this support to interpretation is more than necessary for the reliability expert, both for understanding the problem posed, collecting the relevant input data, understanding the physics of phenomena and interpreting results.

These three facts have doubtlessly accelerated the progress of the system reliability engineering and sciences over the years.

8.2. Challenges linked to reliability?

Reliability is gaining ever growing importance in the industrial world, and even more in the nuclear world. This approach should contribute to optimizing performances and to a continuous improvement. It should also make it possible to detect potential degradations in a system reliability or safety levels so as to remedy them before more serious events occur. One of the aims is to estimate a failure probability. Two major difficulties may then be identified: the validation of the input data and substantially a long and expensive calculation time in industrial-scale cases.

The aims of reliability are numerous:

– making a system – structure – component reliable, avoiding occurrence of a dysfunction or failure in the future;

– providing expected performances through the lifetime (especially the availability of power plants) and fulfilling a public service role (production at the lowest possible cost in complete safety);

– guaranteeing security/safety, seizing all opportunities to strengthen it;

– controlling failures by detecting them in advance and optimizing maintenance interventions;

– lessening potential serious damage to the environment;

– increasing production and its profitability;

– avoid excessive costs or financial losses through a high level of reliability;

– carrying out studies on quality and durability to master ageing;

– improving system's tolerance to failures to improve its robustness and resilience.

The methods presented in this state of the art are very diverse. They have all been applied and some of these applications are presented in this book.

System reliability engineering still has a good future. Many problems remain unsolved. We have identified some main actions, in R&D or in operations, which we believe to be a priority and should be treated over the coming years. These actions are specified in the following section.

8.3. Prospects for future

8.3.1. *Operational feedback data and data quality*

The first perspective is still an operational objective: collecting accurate, good quality and exhaustive data. Data collected more easily now through modern collection methods and stored because of big data computer architectures are not enough. Collection from important situations (failures and events) as it has been done until the present should remain. The redaction of reports on safety, maintenance, human and organizational factors, and perhaps other areas (crisis management, materials, etc.) should be maintained.

Operation experience feedback is strategic. Value and quality of the reliability results are essential for guaranteeing trust. This problem already exists with small-size data sets where reliability experts "manually" qualify the data that they are going to use and distinguish unusual or extreme data using descriptive methods or data classification and discriminant analysis methods. In big data, there will be much more data, big and heterogeneous data, and "manual" validation can no longer be envisaged. Extending these data processing methods to support big data processing seems to us an essential research theme [KEN 14, LAW 17].

Natural language processing (NLP) should help [IMD 13]. Envisaged since the 1990s and then abandoned due to the performance of tools of that period, its use for text processing can be taken up again. That has already been the case in various industries since 2015. We can cite the reference [GAU 18] which illustrates this aspect and its concrete implementation in the hydraulic electricity production sector.

8.3.2. *On system reliability*

Methods currently used in systems reliability are classical dependability (RAMS) methods (failure and event trees). They are enormously useful and will be still in the future. They must not be rejected.

However, more and more, systems are becoming increasingly complex. They are industrial, formed of multiple layers of components that interact in a nonlinear dynamic, with actions and feedback loops. Classical models are no longer sufficient when it is no longer possible to break a system down into independent or isolated sub-systems to study the system's overall behavior. R&D actions should be engaged in, perhaps beginning with a look at compartmental-psychological and biological models of living things [ZWI 06] and simulation models from eliciting expertise. Models of probabilistic networks also seem to be a point of development. One proof comes from Bayesian networks tools that have already been used, effectively, since the start of the 2000s, in risk analysis, in anticipation and in prospect.

Dynamic PSAs are an avenue for research that seems important to us, and which should prove very useful for demonstrating safety, the aim being to break free of the limitations of the classical tools mentioned in Chapter 3. Their benefit is to consider repairs and to precisely reproduce the evolution of physical parameters such as the temperature and pressure of the primary circuit depending on the availability of different cooling methods. Deterministic and thermo-hydraulic physical codes are therefore needed and make it possible to analyze the complete unfolding of an accident. It can easily be understood that calculation costs will be higher, especially as these physical codes will be paired and piloted by programs making it possible to inject random events (failures and repairs) in simulations and to launch a large number of patches using Monte Carlo simulation. Hence, appropriate Monte Carlo acceleration techniques should be developed and computing time should be mastered. We note however that the stakes are worth it. Improvements will be significant since the modeling hypotheses are much more realistic [DOS 18, SHU 20].

The previous tasks in this section are typically R&D tasks. It may be added that other more operational tasks, described in Chapter 3, which are not currently mature due to a lack of applications, since they require testing to demonstrate the potential difficulties and to show their benefit for safety engineers. This involves:

– a modeling innovation, using a Boolean logic Driven Markov Processes (BDMP) model [BOU 03], that makes it possible to gain clarity and precision by replacing the classical, two-model approach using a single BDMP model;

– an innovation in quantification called Initiator and All Barriers (I&AB) [BOU 18], using existing models and improving the relevance of the calculations made.

8.3.3. *On the reliability of structures*

In the field of structural reliability, methods have evolved a great deal since the 1970s. FORM/SORM approximation methods make it possible to provide interesting elements of response, taking account of a much-reduced calculation cost, for many industrial-scale reliability problems in structural mechanics. However, FORM/SORM approximations can result in erroneous assessment of the failure probability. Operational software packages like OpenTURNS exist and are used in several industries. Nevertheless, there are still too few applications. It would make sense to exchange experiences and good practices, diffuse methods and approaches, and produce practical guides for the user [LAN 18]. Besides, industrial and academic bench-marking activities may be of great help.

Essential knowledge for successfully carrying out reliability analyses in structural mechanics is the understanding of the physics of phenomena (applied loads, potential degradations, possible defects, structural behavior up to failure and collapse). A physical equation and a physical interpretation are sources of trust. For phenomena which physical modeling is not sufficiently advanced, it can be useful to calibrate physical models from big data operation feedback or developing metamodels. These are data-driven models issued from machine learning applications. This is an opportunity for understanding and extending physical laws. It allows to update the coefficients of existing physical laws based on observation and to estimate intervals of trust [CIN 20]. These data-driven techniques seem applicable for example to the physical law of degradation (see section 8.3.4).

In addition to advanced simulation methods for estimating failure probabilities, the increasing use of metamodels (such as the polynomial chaos model, the kriging model, etc.) to lower the computing cost linked to advanced numerical codes in structural reliability analysis requires that kind of calibration. Kriging is a meta-model much used in practice. It seeks to approximate a system's response by carrying out a random Gaussian process. Adaptative methods called AK (Active learning and Kriging, [LEL 17], see section 4.1.5.3) have emerged in the past few years, and are developed and used in reliability analysis, particularly for estimating the low probability of failure, or for the reliability of a system or a time-dependent reliability. They consist of a classification method (points of failure/non-failure) based on adaptative kriging, and make it possible to improve the metamodel in zones where the risk of misclassifying points is strong, using Monte Carlo simulations, such as in a classic reliability analysis.

Current applications are also at the root of methodological development needs or the evolution of uncertainty processing tools, such as the processing of system events (overall risk calculations or annual probabilities of failure), methods adapted to some particular cases (numerical integration, conditional Monte Carlo method), sensitivity indices especially for system events, robustness analyses of reliability results vis-à-vis epistemic uncertainties (especially using the info-gap method [BEN 06]) to cite only a few points of current interest.

It seems important to develop these approaches, to establish a comprehensive state-of-the-art of them, and to carry on numerous demonstrative applications, to prove their benefit. The expected gains should be assessed in terms of uncertainties level of the results and computing costs. It is also necessary to determine the difficulties linked to their use and to write practical user-guides.

8.3.4. *On data from big data and the reliability of equipment*

In the domain of component reliability, prospects are also linked to big data. We observe the huge technological development regarding Information and Communication Technologies ([ICT], costly instrumentation and more reliable devices, generalization of connected objects, miniaturization, augmentation of the performances of components controlling means and information collection, increased data storage capacities, etc.). Big data is gradually becoming a reality for the industrial world. Previous studies [MEE 14, HON 18, IMD 20] present these opportunities and show the adaptative use of the classical reliability models, considering covariates and new data (hitherto inaccessible). That is both dynamic, multi-dimensional, voluminous and heterogeneous (time series of operation data and environment measurements taken from a network of sensors, images or films coming from visual inspections using drones, acoustic signals recorded by detectors, textual reports on maintenance interventions, inspection – control data). Making successful use of new information can make it possible to diagnose the state of components' health more accurately and in more individualized fashion, monitor its operational evolution, and assess environmental and equipment maintenance conditions more precisely to predict their effects on reliability; on degradation analysis and degradation kinetics, to prognosticate more accurately the future behavior of equipment (prognostics and health management [PHM]) and its remaining life (RUL). Finally, it allows for adapting the maintenance in consequence, and improving the safety and overall performance of the industrial installations (see section 5.8). It seems necessary to us to multiply applications with greater diversity and complexity, construct experiments (physical and numerical), write practical user-guides, capitalize on information and knowledge, develop operational feedback, gain better knowledge of uncertainties and improving prognosis techniques.

In [LEE 18], authors propose a reliability analysis approach that is conceptually far placed from the current more classical one (presented in Chapter 5). The new approach is building a deep neural network to infer (learn) the lifetime of the studied equipment. Such machine learning algorithms require a large amount of data to be effective, they sometimes present difficulties with interpretation (black box effect), and do not easily allow quantification of the uncertainty linked to the results, or to guarantee trust in the results. However, these authors highlight their ability to integrate nonlinear behavior and to sense information contained in the data that would be difficult to explain using classical probabilistic models.

In our opinion, such an approach will also tend toward an improvement in the understanding of physical phenomena. This means aiding the interpretation of big data results, demonstrating correlations, importance factors, then better constructing and amending physical models, and so better modelling and explaining phenomena. This is a form of hybrid, physical and data-based approach, already envisaged by a shared EDF-Thalès-Total Laboratory in artificial intelligence.

8.3.5. *On the reliability of organizations and activities*

The human and organizational dimensions of reliability are impacting greatly on the reliability of organizations and human activities. Man is the root cause of most major accidents (61.9% in 2012, according to the ARIA database). Although humans make mistakes, they are also a reliability factor in activities (recovering and avoidance, especially in cases of equipment failure).

The first studies on human reliability were made for Wash-1400. Human reliability data has been introduced since the 1970s in PSAs. At the time, little quantitative data were available. Many contested it and did not consider it to be relevant! One operational perspective, it would be better to profit from better informed and more abundant operation feedback, and to update this human reliability data. One might also pursue leadership activities in control room simulations in an international framework (e.g. during testing of the Halden reactor in Norway).

It can be shown that many human reliability analysis methods exist in technical documentation, but, in practice, no method has really imposed itself in the industrial world. Perhaps this is due to a systemic approach or to an unduly detailed orientation that is preventing any progress. The human factor is nevertheless essential for demonstrating safety, or during phases of the lifecycle, for design choices, for operating procedures or organizing maintenance.

The MERMOS [LEB 10] methodology developed from the end of the 1990s is a reference for estimating human reliability; it is used by EDF in PSA. Probabilistic networks that can take into account interactions and uncertainties can also contribute to amply the quantitative analysis [AGU 13]. These probabilistic methods can be used with uncertain variables, whether qualitative or quantitative. It is obviously necessary to model human behavior in its context and in its environment. In this sense, the work of Embrey [EMB 92] seems important to us. Using the influence diagram, it is also possible to model human behavior by considering organizational factors, cognitive factors and variables of context or environment. These modellings were carried out 20 years ago, in the context of maintenance organization. Even though they have limits, such actions should be renewed and developed. We note that probabilistic networks are used in human sciences and in prospective.

Artificial intelligence can also be revealing of human and organizational behavior during operation. Multiple data are now recorded by HUMS systems: data on context, on operation, on monitoring, "manual" data recorded by operators, safety or maintenance reports, published data. It may be imagined that it is possible to build a statistical forecasting model, interpretable (through supervised machine learning) to improve the understanding of human behavior, and to list the explanatory variables of this behavior and their degree of importance. It seems to us that work should be undertaken toward this objective.

It can be stated that human and organizational factors are now better considered from the design stage and in probabilistic safety studies. Progress is significant, as Forester *et al.* [FOR 09] state, where it is postulated that it has become essential to understand and model the cognitive aspects of human behavior and to list the factors that can influence human performance. It concludes that human and organizational factors can be predicted and estimated by their probability. It should nonetheless be recalled that big data is capable of predicting the median of current situations, but not of the unexpected ones. It provides a systemic support for a clearer decision.

Beyond these prospects in human reliability, many fields of progress need to be considered, since they reveal the increasing reliability of activities leadership and maintenance or organizations and risk management (see Chapter 6). One of the breakthroughs in terms of study methodology is to consider the system as sociotechnical and consequently, this change of representation invites us to conduct reliability and safety studies that integrate the technical, human and organizational dimensions. The necessity for integration and interdisciplinarity is always easy to state, but difficult and costly to realize. In addition, these studies should be carried out with better consideration of the weight of the organizational dimensions in the occurrence of sociotechnical failures and immerse ourselves in more systemic modeling, considering feedback loops, the dynamic character nature of some interactions and temporality. R&D studies are rare however, and there remains a great deal of field experience to be crossed for these to be operational and practiced by operators. To this end, networks of actors sensitive to the human and organizational dimensions are a lever for diffusing these approaches, which it would make sense to reinforce. This experience should also serve to diffuse dependability approaches by relying on communities of practice that share their needs and practical difficulties to effectively mobilize the methodological principles specified in this work.

8.4. References

The authors of this book consider these references to be starting points for the perspectives cited in section 8.3.

[AGU 13] AGUIRRE F., SALLAK M., SCHÖN W. *et al.*, "Application of evidential networks in quantitative analysis of railway accidents. Proceedings of the Institute of Mechanical Engineers", *Journal of Risk and Reliability*, vol. 227, no. 4, pp. 368–384, 2013.

[BEN 06] BEN-HAÏM Y., *Info-Gap Decision Theory: Decision Under Severe Uncertainty*, Elsevier, Amsterdam, 2006.

[BOU 03] BOUISSOU M., BON J.-L., "A new formalism that combines advantages of fault-trees and Markov models: Boolean logic driven Markov processes", *Reliability Engineering and System Safety*, vol. 82, no. 2, pp. 149–163, 2003.

[BOU 18] BOUISSOU M., "Extensions of the I&AB method for the reliability assessment of the spent fuel pool of EPR", *Esrel 2018*, Trondheim, 2018.

[CIN 20] CINNELLA P., Data-driven discovery and uncertainty quantification of turbulence models for fluid dynamics, Sciences Sorbonne Université, 2èmes Rencontres chercheurs/ingénieurs "Saisir le mouvement", 2020.

[DOS 18] DOSDA T., BRANDELET J.-Y., BRUNELIÈRE H. *et al.*, *Evaluation Probabiliste de la sûreté dynamique par la méthode des réseaux de Petri*, Congrès λμ 21, Reims, 2018.

[EMB 92] EMBREY D.E., "Incorporating management and organisational factors into probabilistic safety assessment", *Reliability Engineering and System Safety*, vol. 38, pp. 199–208, 1992.

[FOR 09] FORESTER J.A., COOPER S.E., LOIS E. *et al.*, An overview of the evolution of human reliability analysis into the context of probabilistic risk assessment, Sandia Report SAND2008-5085, 2009.

[GAU 18] GAUCHER C., RAYNAL C., URIELI A., *Utilisation du TAL pour la consolidation des données de fiabilités issues des REX*, Congrès λμ21, Reims, 2018.

[HON 18] HONG Y., ZHANG M., MEEKER W.Q., "Big data and reliability applications: the complexity dimension", *Journal of Quality Technology*, vol. 50, no. 2, available at: https://arxiv.org/abs/1803.06206, 2018.

[IMD 13] IMDR, Projet Méthodes d'analyse textuelle pour l'interprétation des REX humains, organisationnels et techniques, Institut pour la Maîtrise des Risques, Cachan, 2013.

[IMD 20] IMDR, Projet IMdR P17-4. Big Data in Reliability, T3 – Impact du Big Data sur l'estimation de la fiabilité d'un composant, d'une fonction ou d'un système, Institut pour la Maîtrise des Risques, Cachan, 2020.

[KEN 14] KENETT R.S., SHMUELI G., "On information quality", *Journal of the Royal Statistical Society, Series A*, vol. 177, no. 1, pp. 3–38, 2014.

[LAN 18] LANNOY A., LEMAIRE M., DELAGE A. *et al.*, *La fiabilité en mécanique – Des méthodes aux applications*, Presses des Mines, Paris, 2018.

[LAW 17] LAWRENCE N.D., *Data Readiness Levels*, Amazon Research Cambridge and University of Sheffield, available at: https://arxiv.org/abs/1705.02245, 2017.

[LEB 10] LE BOT P., "Overview of the MERMOS human reliability analysis method", Idaho Falls, available at: https://secure.inl.gov/isrcs2010/docs/abstracts/LeBot_MERMOS.pdf, 2010.

[LEE 18] LEE C., ZAME W.R., YOON J. *et al.*, "Deephit: a deep learning approach to survival analysis with competing risks", *Thirty-Second AAAI Conference on Artificial Intelligence*, New Orleans, 2018.

[LEL 17] LELIÈVRE N., BEAUREPAIRE P., MATTRAND C. *et al.*, "La famille des méthodes AK pour l'analyse de fiabilité: synthèse des récentes avancées", *23ème Congrès Français de Mécanique*, Lille, 28 August–1 September 2017.

[MEE 14] MEEKER W.Q., HONG Y., "Reliability meets big data: opportunities and challenges", *Quality Engineering*, vol. 26, pp. 102–116, 2014.

[SHU 20] SHUKLA D.K., JOHN ARUL A., "A review of recent dynamic reliability analysis methods and a proposal for a smart component methodology", in: Varde P., Prakash R., Vinod G. (eds), *Reliability, Safety and Hazard Assessment for Risk-Based Technologies*, Lecture Notes in Mechanical Engineering, Springer, Singapore, 2020.

[ZWI 06] ZWIRN H., *Les systèmes complexes – Mathématiques et biologie*, Odile Jacob Sciences, Paris, 2006.

List of Authors

Emmanuel ARDILLON
PRISME Department
EDF R&D
EDF Lab Paris-Chatou
Chatou
France

Marc BOUISSOU
PERICLES Department
EDF R&D
EDF Lab Paris-Saclay
Palaiseau
France

Nicolas DECHY
Department of Human and Organizational Factors of Risk Management
Institut de Radioprotection et de Sûreté Nucléaire (IRSN)
Fontenay-aux-Roses
France

Yves DIEN
CHAOS Association
Meudon
France

Mohamed EID
European Safety, Reliability and Data Association (ESReDA)
European Safety and Reliability Association (ESRA)
Institut pour la Maïtrise des Risques (IMdR)
RiskLyse
Orsay
France

Antoine GRALL
Technical University of Troyes
and
Doctoral school "Science for the Engineer"
France

André LANNOY
European Safety, Reliability and Data Association (ESReDA)
Retour d'EXpérience Technique Sécurité et Sûreté des Structures
Institut pour la Maïtrise des Risques (IMdR)
Cachan
France

Philippe LE POAC
Institut pour la Maîtrise des Risques (IMdR)
Cachan
France

Emmanuel REMY
PRISME Department
EDF R&D
EDF Lab Paris-Chatou
Chatou
France

Jean-François VAUTIER
Direction de la Sûreté Nucléaire et de la Qualité (DSNQ)
Commissariat à l'Énergie Atomique (CEA)
Saclay
France

Index

A, B, C

D, E, F

G, H, I

M, N, O

P, R

S, T, U, V

Other titles from

iSTE

in

Mechanical Engineering and Solid Mechanics

2022

BAYLE Franck
Product Maturity 1: Theoretical Principles and Industrial Applications (Reliability of Multiphysical Systems Set – Volume 12)
Product Maturity 2: Principles and Illustrations (Reliability of Multiphysical Systems Set – Volume 13)

2021

CHALLAMEL Noël, KAPLUNOV Julius, TAKEWAKI Izuru
Modern Trends in Structural and Solid Mechanics 1: Static and Stability
Modern Trends in Structural and Solid Mechanics 2: Vibrations
Modern Trends in Structural and Solid Mechanics 3: Non-deterministic Mechanics

DAHOO Pierre Richard, POUGNET Philippe, EL HAMI Abdelkhalak
Applications and Metrology at Nanometer Scale 1: Smart Materials, Electromagnetic Waves and Uncertainties (Reliability of Multiphysical Systems Set – Volume 9)

Applications and Metrology at Nanometer Scale 2: Measurement Systems, Quantum Engineering and RBDO Method (Reliability of Multiphysical Systems Set – Volume 10)

LEDOUX Michel, EL HAMI Abdelkhalak
Heat Transfer 1: Conduction (Mathematical and Mechanical Engineering Set – Volume 9)
Heat Transfer 2: Radiative transfer (Mathematical and Mechanical Engineering Set – Volume 10)

PLANCHETTE Guy
Cindynics, The Science of Danger: A Wake-up Call (Reliability of Multiphysical Systems Set – Volume 11)

2020

SALENÇON Jean
Elastoplastic Modeling

2019

BAYLE Franck
Reliability of Maintained Systems Subjected to Wear Failure Mechanisms: Theory and Applications
(Reliability of Multiphysical Systems Set – Volume 8)

BEN KAHLA Rabeb, BARKAOUI Abdelwahed, MERZOUKI Tarek
Finite Element Method and Medical Imaging Techniques in Bone Biomechanics (Mathematical and Mechanical Engineering Set – Volume 8)

IONESCU Ioan R., QUEYREAU Sylvain, PICU Catalin R., SALMAN Oguz Umut
Mechanics and Physics of Solids at Micro- and Nano-Scales

LE VAN Anh, BOUZIDI Rabah
Lagrangian Mechanics: An Advanced Analytical Approach

MICHELITSCH Thomas, PÉREZ RIASCOS Alejandro, COLLET Bernard, NOWAKOWSKI Andrzej, NICOLLEAU Franck
Fractional Dynamics on Networks and Lattices

SALENÇON Jean
Viscoelastic Modeling for Structural Analysis

VÉNIZÉLOS Georges, EL HAMI Abdelkhalak
Movement Equations 5: Dynamics of a Set of Solids
(Non-deformable Solid Mechanics Set – Volume 5)

2018

BOREL Michel, VÉNIZÉLOS Georges
Movement Equations 4: Equilibriums and Small Movements
(Non-deformable Solid Mechanics Set – Volume 4)

FROSSARD Etienne
Granular Geomaterials Dissipative Mechanics: Theory and Applications in Civil Engineering

RADI Bouchaib, EL HAMI Abdelkhalak
Advanced Numerical Methods with Matlab® 1: Function Approximation and System Resolution
(Mathematical and Mechanical Engineering SET – Volume 6)
Advanced Numerical Methods with Matlab® 2: Resolution of Nonlinear, Differential and Partial Differential Equations
(Mathematical and Mechanical Engineering SET – Volume 7)

SALENÇON Jean
Virtual Work Approach to Mechanical Modeling

2017

BOREL Michel, VÉNIZÉLOS Georges
Movement Equations 2: Mathematical and Methodological Supplements
(Non-deformable Solid Mechanics Set – Volume 2)
Movement Equations 3: Dynamics and Fundamental Principle
(Non-deformable Solid Mechanics Set – Volume 3)

BOUVET Christophe
Mechanics of Aeronautical Solids, Materials and Structures
Mechanics of Aeronautical Composite Materials

BRANCHERIE Delphine, FEISSEL Pierre, BOUVIER Salima, IBRAHIMBEGOVIC Adnan
From Microstructure Investigations to Multiscale Modeling: Bridging the Gap

CHEBEL-MORELLO Brigitte, NICOD Jean-Marc, VARNIER Christophe
From Prognostics and Health Systems Management to Predictive Maintenance 2: Knowledge, Traceability and Decision
(Reliability of Multiphysical Systems Set – Volume 7)

EL HAMI Abdelkhalak, RADI Bouchaib
Dynamics of Large Structures and Inverse Problems
(Mathematical and Mechanical Engineering Set – Volume 5)
Fluid-Structure Interactions and Uncertainties: Ansys and Fluent Tools
(Reliability of Multiphysical Systems Set – Volume 6)

KHARMANDA Ghias, EL HAMI Abdelkhalak
Biomechanics: Optimization, Uncertainties and Reliability
(Reliability of Multiphysical Systems Set – Volume 5)

LEDOUX Michel, EL HAMI Abdelkhalak
Compressible Flow Propulsion and Digital Approaches in Fluid Mechanics
(Mathematical and Mechanical Engineering Set – Volume 4)
Fluid Mechanics: Analytical Methods
(Mathematical and Mechanical Engineering Set – Volume 3)

MORI Yvon
Mechanical Vibrations: Applications to Equipment

2016

BOREL Michel, VÉNIZÉLOS Georges
Movement Equations 1: Location, Kinematics and Kinetics
(Non-deformable Solid Mechanics Set – Volume 1)

BOYARD Nicolas
Heat Transfer in Polymer Composite Materials

CARDON Alain, ITMI Mhamed
New Autonomous Systems
(Reliability of Multiphysical Systems Set – Volume 1)

DAHOO Pierre Richard, POUGNET Philippe, EL HAMI Abdelkhalak
Nanometer-scale Defect Detection Using Polarized Light
(Reliability of Multiphysical Systems Set – Volume 2)

DE SAXCÉ Géry, VALLÉE Claude
Galilean Mechanics and Thermodynamics of Continua

DORMIEUX Luc, KONDO Djimédo
Micromechanics of Fracture and Damage
(Micromechanics Set – Volume 1)

EL HAMI Abdelkhalak, RADI Bouchaib
Stochastic Dynamics of Structures
(Mathematical and Mechanical Engineering Set – Volume 2)

GOURIVEAU Rafael, MEDJAHER Kamal, ZERHOUNI Noureddine
From Prognostics and Health Systems Management to Predictive Maintenance 1: Monitoring and Prognostics
(Reliability of Multiphysical Systems Set – Volume 4)

KHARMANDA Ghias, EL HAMI Abdelkhalak
Reliability in Biomechanics
(Reliability of Multiphysical Systems Set –Volume 3)

MOLIMARD Jérôme
Experimental Mechanics of Solids and Structures

RADI Bouchaib, EL HAMI Abdelkhalak
Material Forming Processes: Simulation, Drawing, Hydroforming and Additive Manufacturing
(Mathematical and Mechanical Engineering Set – Volume 1)

2015

KARLIČIĆ Danilo, MURMU Tony, ADHIKARI Sondipon, MCCARTHY Michael
Non-local Structural Mechanics

SAB Karam, LEBÉE Arthur
Homogenization of Heterogeneous Thin and Thick Plates

2014

ATANACKOVIC M. Teodor, PILIPOVIC Stevan, STANKOVIC Bogoljub, ZORICA Dusan
Fractional Calculus with Applications in Mechanics: Vibrations and Diffusion Processes
Fractional Calculus with Applications in Mechanics: Wave Propagation, Impact and Variational Principles

CIBLAC Thierry, MOREL Jean-Claude
Sustainable Masonry: Stability and Behavior of Structures

ILANKO Sinniah, MONTERRUBIO Luis E., MOCHIDA Yusuke
The Rayleigh−Ritz Method for Structural Analysis

LALANNE Christian
Mechanical Vibration and Shock Analysis – 5-volume series – 3rd edition
Sinusoidal Vibration – Volume 1
Mechanical Shock – Volume 2
Random Vibration – Volume 3
Fatigue Damage – Volume 4
Specification Development – Volume 5

LEMAIRE Maurice
Uncertainty and Mechanics

2013

ADHIKARI Sondipon
Structural Dynamic Analysis with Generalized Damping Models: Analysis

ADHIKARI Sondipon
Structural Dynamic Analysis with Generalized Damping Models: Identification

BAILLY Patrice
Materials and Structures under Shock and Impact

BASTIEN Jérôme, BERNARDIN Frédéric, LAMARQUE Claude-Henri
Non-smooth Deterministic or Stochastic Discrete Dynamical Systems: Applications to Models with Friction or Impact

EL HAMI Abdelkhalak, RADI Bouchaib
Uncertainty and Optimization in Structural Mechanics

KIRILLOV Oleg N., PELINOVSKY Dmitry E.
Nonlinear Physical Systems: Spectral Analysis, Stability and Bifurcations

LUONGO Angelo, ZULLI Daniele
Mathematical Models of Beams and Cables

SALENÇON Jean
Yield Design

2012

DAVIM J. Paulo
Mechanical Engineering Education

DUPEUX Michel, BRACCINI Muriel
Mechanics of Solid Interfaces

ELISHAKOFF Isaac *et al.*
Carbon Nanotubes and Nanosensors: Vibration, Buckling and Ballistic Impact

GRÉDIAC Michel, HILD François
Full-Field Measurements and Identification in Solid Mechanics

GROUS Ammar
Fracture Mechanics – 3-volume series
Analysis of Reliability and Quality Control – Volume 1
Applied Reliability – Volume 2
Applied Quality Control – Volume 3

RECHO Naman
Fracture Mechanics and Crack Growth

2011

KRYSINSKI Tomasz, MALBURET François
Mechanical Instability

SOUSTELLE Michel
An Introduction to Chemical Kinetics

2010

BREITKOPF Piotr, FILOMENO COELHO Rajan
Multidisciplinary Design Optimization in Computational Mechanics

DAVIM J. Paulo
Biotribolgy

PAULTRE Patrick
Dynamics of Structures

SOUSTELLE Michel
Handbook of Heterogenous Kinetics

2009

BERLIOZ Alain, TROMPETTE Philippe
Solid Mechanics using the Finite Element Method

LEMAIRE Maurice
Structural Reliability

2007

GIRARD Alain, ROY Nicolas
Structural Dynamics in Industry

GUINEBRETIÈRE René
X-ray Diffraction by Polycrystalline Materials

KRYSINSKI Tomasz, MALBURET François
Mechanical Vibrations

KUNDU Tribikram
Advanced Ultrasonic Methods for Material and Structure Inspection

SIH George C. *et al.*
Particle and Continuum Aspects of Mesomechanics

Printed and bound by CPI Group (UK) Ltd, Croydon, CR0 4YY
16/09/2026
14963088-0005